Financial Dynamics

A System for Valuing Technology Companies

Financial Dynamics

A System for Valuing Technology Companies

Chris Westland

John Wiley & Sons (Asia) Pte Ltd

Published in 2003 by John Wiley & Sons (Asia) Pte Ltd
2 Clementi Loop #02-01, Singapore 129809

■ Other Wiley Editorial Offices:
John Wiley & Sons, Inc., 111 River Street, Hoboken, NJ 07030, USA
John Wiley & Sons Ltd, The Atrium, Southern Gate, Chichester, PO19 8SQ, England
John Wiley & Sons (Canada) Ltd, 22 Worcester Road, Rexdale, Ontario M9W 1L1, Canada
John Wiley & Sons Australia Ltd, PO Box 174, North Ryde, NSW 2113, Australia
Wiley – VCH, Boschstrasse 12, 69469 Weinheim, Germany

■ Library of Congress Cataloging in Publication Data
ISBN 0-470-82111-6

Typeset in 11/14 points, Times New Roman by Paul Lim.
Printed in Singapore by Saik Wah Press Pte Ltd.
10 9 8 7 6 5 4 3 2 1

To Randa, Sumar, Dina, and Brendon

Contents

Acknowledgments

I owe many debts – both personal and intellectual – to my professional colleagues, business acquaintances, and personal friends who have helped with and commented on the material in this and my prior book, *Valuing Technology: The New Science of Wealth in the Knowledge Economy*. Any reassessment of received wisdom is both difficult and holds risks, which I ultimately must bear. Eventually, the ideas and methods that are laid out in the ensuing pages are my own responsibility.

But I was influenced by many people along the way.

I wish to thank, in particular, my colleagues both at Nanyang Technological University in Singapore for their many suggestions and for providing me with the freedom and support during my sabbatical year to complete this book, and at the University of Science and Technology in Hong Kong for their many comments which have helped shape the work you see presented here. Sam Moon of DNM Strategies originally suggested these books and has provided immense support in promoting the ideas contained therein.

I owe intellectual debts to George Gilder, who first articulated many of the phenomena which have been incorporated into the Financial Dynamics model; and to Hal Varian; whose classes at the University of Michigan introduced me to many of the useful insights into the economics of the knowledge economy which make this work possible, Jan Kmenta and Bruce Hill also at the University of Michigan, instructed me on many of the tools required to tackle this work. Antithetically, I have to credit

Bill Kinney and Joel Demski who, though perhaps unwittingly, convinced me of the importance of a radical reassessment in financial accounting. Barry Boehm and I developed some of the techniques presented here in our work on the Cocomo software cost estimation program; David Brooks furthered the refinement of these techniques in helping me apply these to Microsoft's international business operations. In the process, I have benefited from presentations of this material in seminars at the University of Michigan, USC, UCLA, New York University, Tulane, Baruch, and other institutions, and from feedback from my colleagues in information systems and economics: Chris Kemerer, Eric Brynjolfsonn, Avi Seidmann, Haim Mendelson, Rajiv Banker, Baruch Lev, Ted Clark, Rob Kaufmann, David Webb, and Gary Biddle.

I am certain that there are other colleagues whose contributions to this work I have overlooked, and I ask them to please accept my apologies for the omission.

Nor would any of this have been possible without the support, perseverance, and hard work of the staff at Wiley. I wish especially to thank Nick Wallwork, who has supported this project from the start, and Malar Manoharan and Robyn Flemming for their insights, suggestions and editing of my manuscripts.

And finally, I owe more than I can say to the support and love of my family.

Preface

Technology businesses today are unlike anything we have seen in history. They are fluid, reactive, directed by individuals who are managers, workers, consumers, and investors, and greased by the greatest communication system in history. Wealth in the form of natural resources counts for less. Information races in and out of every nook and cranny. Actions get weighted at every desk and in every home. Phenomenal resources are redirected in a flash. Over the next decades, our technology-laden knowledge economy will grow even more pervasive.

The knowledge economy is fundamentally different than the industrial and agricultural economies that preceded it, a fact emphasized in two devastating financial crises that have occurred in just the past two decades. Both crises made apparent the crucial shortcomings in financial accounting, discounted cash flow (DCF), and other valuation methods when applied to knowledge-intensive business.

On "Black Monday" (October 19, 1987), the Dow Jones Industrial Average tumbled 22%, the culmination of fiddles by microcomputer-empowered bond salesmen playing on finance and accounting's inability to assess the value of high yield risk. The dotcom and telecommunications debacle of the summer of 2000 saw US$8 trillion evaporate after years of irrational exuberance of firms rich in ideas and not much else. These two crises were ugly reminders that our capital markets simply cannot function without a language that can accurately and transparently describe what traders are buying and selling.

Unfortunately, our current "language of value," the vernacular of financial accounting, has grown inarticulate in the past two decades. Fraud makes headlines; errors, omissions, misunderstandings, and misguided strategies present – in dollars lost – much bigger problems. Fraud is intentional and dramatic; poor accounting is prosaic and often the product of ignorance, but on the whole is significantly more damaging. When dollars are lost, jobs are lost as firms go bust, pensions shrivel, and investors become wary of managers, accountants, brokers, and nearly everyone else associated with business. Indeed, such a wave of cynicism has been the consequence of the dotcom and telecommunications meltdowns. The harm that this does to competent businessmen is enormous. Poor accounting has created a market for lemons, where moral hazard and adverse selection captivate the market, dragging down the good along with the bad. The solution: better valuation models and more transparency.

Technology itself has a role to play in delivering the transparency needed to combat fraud and mistrust in a free market. Better information technology systems alone cannot stop fraud or people with their mind set on perpetrating it. But by automating the process, you reduce the risk of human error and human interference, both of which have always greatly exceeded firms' and markets' exposure to risk. The Financial Executives International, in its March 2002 report entitled *Observations and Recommendations for Improving Financial Management, Reporting and Corporate Governance*, concluded that companies must modernize their reporting and move toward voluntary disclosure of additional management performance indicators using real-time Web-based reporting mechanisms. The alternative is a continued erosion of confidence in financial reports, and waning of participation in financial markets.

This book synthesizes new research and current practice into a comprehensive model for corporate valuation in knowledge-intensive fields – what we have come to call "technology businesses." The specific focus of the "language of value" reflects the economic importance of specific business models, the implementation of vertical markets, and core competencies. At the root of this language is the quest for better ways to track the performance of businesses when investments are primarily in intangible assets. Such businesses now contribute over 80% of GDP in developed economies. They include IT powerhouses such as

Microsoft and Cisco, which are being joined by knowledge firms in biotechnology (such as Pfizer and Pharmacia), and in the near future, nanotechnology.

It is my sincere hope that the Financial Dynamics model finds widespread application to valuation, but I will feel that this book has been successful if it simply furthers debate on the options available to accounting and valuation in meeting the challenges of 21st century business.

Chris Westland
Hong Kong and Singapore
May 2003

CHAPTER 1

The Language of Value

In our post-industrial economy, our accounting system doesn't do a good job of describing any modern company.

– Steven Butler, chairman of KPMG, WorldCom's auditor[1]

Our financial reporting model is broken.

– Joseph Berardino, chairman of Arthur Andersen, responsible for WorldCom and Enron's audits[2]

CONFESSIONS OF A CERTIFIED PUBLIC ACCOUNTANT

The excesses of the 1990s' dotcom bubble were not pretty. Seduced by Internet mania and the nirvana of the so-called new economy, investors piled into companies with silly business plans and negative profits. The post-mortem has brought a chorus of demands for retribution, compensation, and change. Arthur Andersen, one of the world's great accounting and consulting firms, collapsed under allegations of practices that were clearly "unprofessional" and "troubling" (in the stated judgment of the firm's own partners). It highlighted an incestuous network of conflicts of interest, shoddy auditing, market-sensitive news releases, and other "disclosure" problems enveloping Wall Street and America's capital markets. Predictable finger-pointing pitted accountants against Wall Street in court and in the news.

What is certain is that these two disparate fraternities shouldered most of the responsibility for corporate mis-valuation during the dotcom and telecoms feeding frenzies. What is not certain is why this happened, or what can be done in the future to prevent a remake of the dotcom bubble. The investment community, with hyperactive analysis and seductive sales pitches, promoted equity investments using valuations that always seemed on their way up. The accounting profession, charged since the 1930s with keeping the investment community in check, provided their moral conscience. Well, that's the theory.

The priesthood of accounting – taciturn, schooled in their labyrinthine dogma, and versed in the litany of audit – is the voice of probity in the temples of industry. Not far away, casting their lots on Wall Street, are the sinners – garrulous traders, cheered on by celebrity analysts. The forces of iniquity are just too hard not to like. Shrewd manipulation of dotcom stocks netted them fortunes, leading to a fashionable greening of Wall Street.[3]

Curiously enough, the same language is spoken in both the temples and the Street – the language of financial accounting.

The priests are empowered to draw up lists of sins and are revered for the special talents they display. Sinners are not expected to avoid sins; they need only confess their errors openly. An annual audit – the plenary indulgence expunging a year of sins and pronouncing financial rectitude– is sought after and sold at top dollar.

But all is not well. The priesthood is saddled with a medieval language which daily grows less relevant to a technology-laden market. Perhaps the priesthood is so ineffective in restraining sin because the sinners themselves cannot tell when they are sinning.

I began thinking about these problems in the 1970s as a certified public accountant (CPA) employed at the accounting firm of Touche Ross.[4] Our offices, in the financial center of Chicago, were shared by corporate financial analysts and investment bankers. When our paths crossed, as they sometimes did on elevators or during lunch, I was perplexed by the fact that the same language was used in both places. Even more amazing was the occasional transmogrification of the highest of high priests in our accounting firm. After leaving the audit firm for a corporate job, they would often metamorphose into particularly wanton sinners. As I struggled against these contrasts, I found that, for all the purported rigor and formalism of our shared language, that language was not, in truth, all it was reputed to be.

I suppose, as a CPA, that I am technically a member of the priesthood, though no doubt that priesthood would consider me apostate. For the past 30 years, I have concentrated on information technology and the management of technology. I have also grown increasingly skeptical of the efficacy of accounting's dated liturgy. Too many features of 21st-century commerce are simply inexpressible in the language of accounting.

The casinos of investment are distanced from the cloisters of accounting on purpose. Accountants were engaged in the United States after 1934 to rein in the unbridled speculation that tarnished the reputation of U.S. securities markets in 1929. The accounting discipline – intellectually defined nearly 600 years earlier by the Italian monk Luca Pacioli – offered a puritanical prescription for such intemperate speculation.

Our present language of accounting evolved from business conventions used in the 14th-century Venetian merchant marine, and in the 16th-century English manorial system for accounting for crops. The fact that neither of these businesses survives in its historical form is sure to cast suspicions that, as the language got old, it was unable to express the needs of a complex and evolving world.

The opinion that the language of financial accounting is precise, rigorous, and objective is held by an embarrassingly large share of both communities. Accountants argue that their numbers are "objective" and "consistent," and thus are as close to the truth as is possible; analysts make similar claims about their predictions. Yet no less an authority than Harvey Pitt, former chairman of the U.S. Securities and Exchange Commission (SEC), observed that "there is no true number in accounting, and if there were, auditors would be the last people to find it."[5] Only charlatans would claim to speak a language in which only the truth is ever told – and "truth" in valuation is as much a matter of degree as intent.

Since there is no such thing as a faultless account valuation, one might conclude that assessments of corporate value without an estimate of the error in that value also must be the work of charlatans. Yet, this is the world of the accounting priesthood. Account valuations are stated with precision to the penny, and any assessment of error in their accounts – the "materiality" – is carefully guarded and reported only under coercion from the courts.

My current book attempts to bridge the gap between the priests and the sinners. One of the outcomes of this effort is an appreciation of why certain sins are unavoidable, or even desirable. A new sin does emerge, however. It is a sin not to know why you are sinning. Pointless sin must be avoided. (More about this in Chapter 11.)

The broad gap between accounting practice, on the one hand, and investment, on the other, might be expected to create tensions. In fact, a calm exists between the priesthood and the legion of inveterate sinner-analysts – an equilibrium lubricated by generous salaries and even more generous commissions.

The sinners on the Street may indeed be sinning because they have no way of understanding the liturgy. Can unavoidable sins truly be sins, after all? Mr. Pitt seems not to think so, remarking that

> No investor – certainly not any ordinary investor – can read these [financial reports] in a way that's useful. An investor can't know what's been left out, why it's left out or how it compares with other companies' calculations. The quality of financial reporting leaves a great deal to be desired. Even if there were no issuer problems, the information is not timely – it's stale when investors get it. Worse, it's impenetrable.[6]

Or, as it was put bluntly by Lehman Brothers accounting expert Robert Willens: "I don't know anyone who uses GAAP net income anymore for anything."

Being a CPA myself, Mr. Willens's and Mr. Pitt's criticisms clearly hit a nerve. For nearly 30 years, I have observed my colleagues both in academe and in practice. I analyzed in my previous book, *Valuing Technology*, my understanding of what it is that both the sinners and the priesthood were up to – and more importantly, how both have grown totally ineffective in addressing the dictates of the knowledge economy.

As this book goes to press, corporate America is in the process of assessing the consequences of the *Sarbanes-Oxley Act*, which is designed to hold accountants more accountable (along with executives and securities firms). At a minimum, this Act, by subjecting auditing and financial statements to more scrutiny (and thus lawsuits), should encourage accountants to focus on getting the accounting right, since it has banned them from eight services that used to comprise most of their earnings – most notably, IT consulting. This should also help IT consultants to more effectively design accounting systems, unencumbered by the 600-year-old traditions of GAAP (though, of course, they will need to tread carefully in this terrain).

Financial Dynamics is, ultimately, about language. It is about how, at the most fundamental level, we describe and summarize the economic information that is ultimately needed for business decision-making. More than that, it presents my attempt to construct a rigorous language with which to measure the particular types of "value" produced in the knowledge economy. In the process, I have borrowed from the modern languages of statistics, engineering, and corporate strategy. I have also borrowed freely from the respective languages of accounting and investment, but have studiously avoided co-opting their rituals, incantations, and dogma. I believe that the new *language of value* described in this book can effectively capture the myriad complexities, subtleties, and peculiarities of business in the knowledge economy.

THE LAST GUILD

The accounting profession that arose after the 1930s grew into a latter-day guild. The cost of entry to the medieval guilds was indenture and apprenticeship; 20th-century accounting required extra schooling and industry practice prior to certification. Guild apprenticeship ensured consistency and quality at the expense of flexibility and innovation. Such controls were not a problem while costs, technology, and products remained static. But since technology has accelerated, such guilds have found themselves unable to compete on cost and quality, and new businesses have developed to usurp their power.

Guilds have a long history, and at one time provided valuable services to their members and to society at large. Merchant and craft guilds[7] arose on the European continent in the 11th century. Craft guilds included in their membership all those engaged in any particular craft, and monopolized the making and selling of a particular product within the cities in which they were organized. They obliged members to stand by each other in legal disputes in which any might engage.

Accounting and writing have been inextricably entwined throughout history. Most scripts grew out of early systems for keeping track of commerce. Some scripts, such as Linear B, are thought to have been used only for commercial accounting. Scribes in medieval Europe were largely confined to the scriptoria, or writing rooms, of the monasteries; outside, the majority of the population was illiterate. Later in the Middle Ages the universities supervised the preparation of textbooks and literary works, and also prescribed the rates at which the books were to be sold or leased. It is not by accident that the definitive medieval treatise on accounting was scribed by the Venetian friar and university academic Luca Pacioli in his 600-page tome, *Summa de aritmetica geometria proportioni et proporionalita* (The Collected Knowledge of Arithmetic, Geometry, Proportion and Proportionality). The work was widely read, and became one of the first books printed on the Gutenberg press.

Guilds of scribes (from the Latin scribere, "to write") represent perhaps the most ancient of guilds. They served as copyists, editors, and interpreters of the Bible and of the law. In ancient and medieval cities, they plied their services to illiterate farmers and craftsmen who would pay for a permanent, legal record of transactions. Their traditions continue to this day in the latter-day professional accounting guild.

By the middle of the 12th century, craft guilds had been established throughout the commercial world. A guild received from government the right to control all practice of its trade – just as, today, accountants must be licenced and certified if they are to practice. The members of a guild were divided into three classes: masters, apprentices, and journeymen. The master was a small-scale proprietor, much like the partner-owner of a public accounting firm. After an apprentice completed his training (today this involves a university degree in accounting, followed by another year of specialized courses, a CPA examination, and a one- or two-year stint of service at a public accounting firm), he would become a journeyman paid a fixed wage for his labor. In time, a journeyman might become a master (partner). Because it was to the advantage of those who were already masters not to increase their own number, the conditions under which a journeyman might become a master were always made difficult. The latter restraint is maintained in large accounting firms today, where only one out of hundreds of "journeymen" are ultimately invited into the partnership. Guilds were central to the life of the medieval city, aiding members by protecting them from competition. The guild permitted no goods from other cities to be imported for sale in its own city, and within the city decreed how many people were to be employed in each shop, the number of tools to be used, the hours of labor, and the prices the master could charge for his finished goods. The guild enforced its rules by constant and close supervision of the shops and through peer reviews. No master was permitted to advertise his goods, lest he attract more business than another master. Any improvement in the technique of production that might enable one shop to produce goods more quickly and cheaply than another was likewise prohibited.

The ultimate cause of the guild system's demise was the Industrial Revolution. This new economic system stressed large-scale production of goods, competition for markets between producers, and wide distribution of goods. Inasmuch as the guild was inclined against all three principles, capitalists generally established their shops in centers where no craft guilds existed.

Guilds today have all but been relegated to the rubbish heap of history, except for the odd anachronism. Typically, these are guilds in a profession whose products and services are seldom in contact with the common man; those which may have remained largely untouched by the Industrial Revolution, or the Information Revolution that followed. Notable among these anachronisms is the accounting profession.

The main players in accounting have been the so-called Big 5 accounting firms (since Arthur Andersen's demise, facetiously called the Final 4). These are accounting firms mostly in name. In 2001, these firms received several times their income from accounting-related services from management consulting, mainly in building corporate information systems. The rectitude of the audit side of these firms often clashed with the evangelical prose of their colleagues in management consulting. The conflicts of interest are palpable, and – post-bubble – the topic of much hand-wringing by an alphabet soup – SEC, NASD (National Association of Securities Dealers), FASB (Financial Accounting Standards Board) – of organizations charged with ensuring that accountants do their jobs right.

Unfortunately, neither side offers accurate predictions of future wealth generation nor do they provide help where corporate value is primarily a function of future cash flows or income. There have been calls in the past for a new language and new approach to valuation, but the task is very difficult. There is growing interest, in regulatory bodies such as the U.S. SEC, in generating a second "parallel" set of accounting statements that accurately reflect "operating income." Other initiatives have come under rubrics such as "fair value accounting."[8] This in itself has proved to be controversial, as are all new ideas that threaten the power base of established communities.

ACCOUNTING IN THE AGE OF UNCERTAINTY

The Old Guard of accounting clung to some rather strange and charming habits, arguing over the teleology of "true income" and the correct pronunciation of "comptroller". They treated accounting as a branch of religion. Their lives were premised on an elitist conviction that they were above the fray: possessing a special knowledge; observing a strict code of conduct without ever having to say what it was.

In the 1950s, accountants sought the holy grail of "true income" through an axiomatic system based on a fixed set of accounting principles that would have been worthy of Aristotle. The axiomatic approach to accounting sought objective "facts" (usually documented historical costs) which could translate, using stock formulas, its theoretical "principles," and the "entities" with which they dealt, into the "essential value" of the firm. As often with such quests, the goal was illusory. "Value," like beauty, lies in the eye of the beholder. Different accounting systems are needed for different purposes.

Perhaps this should not have been a surprise for the accountants. The accounting profession has typically focused on the needs of shareholders. Banks, on the other hand, have looked to credit rating agencies. John Moody invented credit ratings in 1909, when he published the *Manual of Railroad Securities*, which rated 200 railroads and their securities. The Standard Co. began grading bonds in 1916. Poor's and Fitch followed in the 1920s. Poor's and Standard merged in 1941. Inside the growing industrial firms – automobiles, ships, petroleum – at the turn of the century were the cost accountants, whose clientele were corporate managers. Each group imposed their own interests on "value" – credit raters on ability to repay debt; cost accountants on the dynamics of the production process.

No one really questioned accounting's axiomatic tradition until, in the Reaganesque 1980s, accounting (the profession) succumbed to twin American instincts: to democratize and to commercialize (the same thing, in the thinking of the Old Guard). Restructuring was set in motion when the American Institute of Certified Public Accountants (AICPA) repealed its Code of Professional Conduct's sections 502-1, 502-3, and 502-4.

Accounting partners began to receive calls from "consultants" who wanted to help them learn how to steal clients and staff from other firms – a notion that would have been unthinkable to the Old Guard. The minute the market intruded, the old prestige began to seep out of the profession. Wrapped up as their identities were in tradition and jealous of the power of the market, it is not surprising that the Old Guard abandoned accounting's new business model *en masse*. Abolition of these rules, which had prevented auditors from advertising and solicitation, subsequently opened them up to the full blast of market forces. Within the decade, the profession had replaced many of its elders with market-savvy operators. By the end of the Reagan administration, the profession had been transformed – and not necessarily for the better. These New Professionals adopted the dignified facade of the Old Guard, but suffered from the disintegration of the inward allegiances and loyalties that had made the Old Guard truly "professional."

This new generation of partners redefined the "profession" of auditing as a loss-leader for their computer consulting business – a sure mark of the rise of the information economy if there ever was one. Auditing grew less profitable, and more risky. Between 1985 and 2000, audit firms twice saw the industry's contingent liabilities from lawsuits exceed the combined net worth of all the accounting firms in the United States.

The major users of accounting information – the financial analysts at funds, banks, and so forth – have also seen their industries transformed. Online stock trading, which captured close to 40% of market transactions in the latter half of the 1990s, and online banking, which accelerated the shakeout and mergers of retail banks throughout the 1990s, are just two of the impacts of information technology that have dramatically transformed the job of the financial analyst. The rise of Internet stock trading and the availability for free of reports that previously sold for thousands of dollars have forced analysts to seek revenue elsewhere in order to justify their employment. In consequence, they have evolved from the role of financial journalists to market-savvy propagandists for the products and clients of the securities firms.

While both auditing and analysis were transforming, business was growing more complex and dependent on information technology while assets were growing intangible and ephemeral. From the point of view of investors, bondholders, and managers, we had entered an age of uncertainty. Accountants were supposed to dispel the uncertainty. Instead, they contributed to it by continuing to converse in an antiquated language incapable of expressing the constituents of the knowledge economy.

Accountants and analysts have little freedom to change either their industry's current structure or the realities of 21st-century business. But accountants can learn from other disciplines that have suffered similar crises of uncertainty due to the explosion of learning in the 20th century. Physics, for example, has had to cope with its transition from the axiomatic Newtonian system, to relativity (and the discarding of an absolute spatial reference position) and quantum mechanics (where Heisenberg uncertainty makes momentum and position simultaneously uncertain). Physics solved its problem through systems that allowed flexible perspectives to serve different parties – for example, relativity considers how physical reality appears in local reference frames. Physicists also separate interpretation from method in assessing their reality, a distinction made by neither Newtonian nor Aristotelian physics. For example, Schrödinger's wave equation in atomic physics was easy to use but difficult to interpret and impossible to observe, while the equivalent physics according to Heisenberg had clear interpretation, but was difficult mathematically.[9] Lest I be accused of succumbing to "physics envy," let me say that similar developments mark many of the natural sciences in the 20th century, though perhaps physicists have been better at articulating and popularizing the transformation of their science.

The important point – one that promotes both vigorous debate and improved understanding – is that scientists in the age of uncertainty have learned to eschew solely axiomatic depictions of reality, or searches for absolute essences and truth. Instead, they lay out their assumptions for all to see, inviting debate. They conclude and predict, inviting replication of experiments, alternative interpretations, and evolution of their models. Good science today is less about finding absolute truths, than it is about reporting what we know and what we believe, then vigorously debating it in order to improve on it. Such an approach promotes more complete reporting of all the information, and less frivolous pro-forma conjecturing. This book is my attempt to suggest ways in which financial analysis and accounting can evolve to embrace the age of uncertainty as effectively as physics and the natural sciences.

It may be a strange time to say this, just when the world is clamoring for truth in accounting. But what companies really need is greater tolerance of uncertainty. They should be allowed to assign dollar values to intangible assets such as patents, brands, and customer lists. With a better grasp of invisible assets, investors can make wiser decisions.

As the embodiment of ideas, intangibles are what drive growth in an information economy. Baruch Lev, an accounting professor at New York University, reckons that in the late 1990s, businesses invested approximately US$1 trillion per year in such assets. Taken together, intangibles comprise well over half the market value of public companies. Yet, with statements prepared under U.S. generally accepted accounting principles (GAAP), investors can only guess at their value or makeup. Current reporting practices "fail to provide adequate guidance to managers, investors, or public policymakers," says Lev.[10] True to form, Halsey Bullen, senior manager of a project to study intangibles at the FASB, has forcefully argued that the place to record intangibles is "in the footnotes" – not very useful when it comes to ascertaining profit and share value.

Embracing uncertainty will make the job of ascertaining risk an integral part of accounting. Accountants and auditors are in a good position to assess risk, because they have access to proprietary information that investors, banks, and securities firms do not. Accountants must consider three realms of risk: (1) risk from business operations; (2) risk from management's strategy and the way this plays out in the supplier and customer markets; and (3) risk in capital markets, which provide some of the financing for business operations (the rest being internally generated or from lending). The firm's own business operations are under control of management; vendor and customer markets set the environment in which managers make their decisions; and capital markets are subject to a variety of subjective biases. Many big banks already use quantitative models to assess how much capital they need to set aside against portfolios of marketable securities. These "value at risk" (VAR) models typically measure the most the firm could lose in a day, judging by past performance, but they tend to underestimate the frequency with which really bad days occur. There have been half a dozen "perfect storms" in the market in the past decade, during which VAR calculations proved useless in predicting losses. Stress-testing portfolios against imaginary perfect storms remedies some of the weaknesses. But modeling credit risk in this way is much harder, not least because data about past credit performance are scarce.

Government regulation of risk has expanded in three distinct phases in the United States, paralleling the growth in complexity of financial management. In the 19th century, the government typically intervened in order to assist businessmen in the development of an industrial economy through measures such as bankruptcy laws and to reduce risk for entrepreneurs. In the first half of the 20th century, the provision of greater income security for workers and pensioners was promoted, starting with the presidency of Theodore Roosevelt.[11] After World War II, government extended its role as risk manager still further to help consumers act as insurer of last resort for disaster victims and to remedy environmental blight. The characteristic form of modern risk management has been for the government to assume catastrophic risk itself (and consequently spread it across the taxpaying public).

WHY YOU SHOULD READ THIS BOOK

When John Wiley & Sons published *Valuing Technology* – the predecessor to this book – I included several caveats. For one, I noted that I did not have an exact solution for valuing knowledge economy companies, because they operated at a new level of complexity. I instead was intent on studying why there were such discrepancies in the valuations of knowledge-intensive companies.

I purposely limited the scope of *Valuing Technology* to problem articulation. The challenges raised by the emergent knowledge industries – Internet, genomics, telemedicine, robotics, and so forth – were significant. By the time of the dot-com crash there was a widespread awareness of accounting's inadequacies. Even among the large and stable firms of the Fortune 500, market valuations average three to six times accountants' book valuations.

The problem of computing accurate valuations has preoccupied me since then. Whatever the errors and omissions of that book may be, I liked *Valuing Technology*. However, I couldn't be comfortable with the loose ends it contained. This book represents the "sequel." It is not a retread of the first book's topics; in fact, there is very little overlap between the two. It is, instead, my answer to the question, "What is the 'value' of a knowledge-intensive company?"

Financial Dynamics redefines our 600-year-old tradition of financial accounting so as to address the dictates of the knowledge economy. This book describes how subjective knowledge about the structure and operations of a company can be legitimately integrated with the limited but objective information contained in historical financial reports. The essential ingredients are judgment and purpose, which jointly determine where "value" – however we choose to define it – resides, and how effective is our business model in generating it. Without judgment and purpose, corporate valuation is at best a fishing expedition, and at worst, a scam. The product of the "valuation" will itself have value that is difficult or impossible to assess.

Before the reengineering vogue of the early 1990s, the question of business models was seldom raised outside of the information systems community. Up until that time it was commonly assumed that labor, capital, and materials were combined in any business in a fashion that was uniquely determined by the product and its markets. Reengineering was first promoted as the Semantic Data Model and focused on the structure of business operations; only later did it evolve into the prescriptive and holistic reengineering model promoted in the 1990s. The growth of the dotcoms in the late 1990s thrust a variety of complex and creative business structures on to the investment scene, which analysts were required to defend through justifying the value created by the business model.

Information technology has shifted the nexus of business away from labor, capital, and raw materials (which at one time were the major forces in the political economy) and toward the intellectual processes that extract value from these inputs. Because of this, investors want clear delineations of the "business model" on which value is generated. The bulk of innovation in Financial Dynamics centers on constructing valuations around firms' business models.

WHAT THIS BOOK IS NOT

This book is not about stock-picking, portfolio theory, or "beating the market." Though this is possibly one of the uses to which Financial Dynamics may be employed, it is not my main interest in this work. Rather, the focus of the book is on investigating the failures in financial accounting, and providing better alternatives for financial analysts and investors who use the products of accounting.

Although the approach has merit for stock valuation, there are technical aspects of financial markets which are intentionally left beyond the scope of this book – liquidity, trading biases, dealer- versus order-driven markets, and so forth. For the interested, I would recommend reading Lo and MacKinlay's *A Non-Random Walk Down Wall Street* (Princeton, N.J.: Princeton University Press, 1999), along with Campbell, Lo, and MacKinlay's *The Econometrics of Financial Markets* (Princeton, N.J.: Princeton University Press, 1997), which together provide a thorough, objective, and learned survey of the issues that need to be considered in market valuations.

The current book addresses only a small part of portfolio choice. There are at least three main components of stock-market prices: (1) the fundamental value of the firm being traded; (2) the market demand; and (3) feedback effects. The first, I think, can be computed relatively accurately by improving existing sources of data and by borrowing methods from econometric forecasting with the right sort of statistical characteristics. It is this component that is the focus of the current book. There is growing interest in non-GAAP alternatives as credit reporting and boutique metrics looking for better ways to compute the worth of firms. It is not by pure chance that the most successful stock-pickers – pundits like Warren Buffett and Peter Lynch – have advised not to focus on the financial statements, but to take a walk through their retail outlets, shop floors, and the gritty minutiae of daily operations to find their real worth. Financial Dynamics demands this detailed knowledge as a systematic input to the method.

The market demand curve, our second component, results from the combination of liquidity, hype, and other factors specific to the stock instrument and its market. Widely traded securities will be on the flat part of the curve with a price based on fundamental value; thinly traded issues will be on a steep (speculative) side of the demand curve. Where anyone with a server can set up a stock market, we cannot retain the luxury of assuming that stock markets efficiently incorporate fundamental information into their prices. The situation is perhaps most succinctly captured by Massachusetts Institute of Technology economics professor Andrew Lo, who observed that

> … people who cling to the notion of an efficient market suffer from a peculiar psychological disorder known as "physics envy." We

> would love to have three laws that explain 99% of economic behavior; instead, we have about 99 laws that explain maybe 3% of economic behavior. Nevertheless, we like to talk as if we are dealing with physical phenomena.[13]

Lo suggests that we would be better off treating market efficiency in terms of relative efficiency, like that of a gasoline engine where we think in terms of a percentage efficiency – for instance, an engine is 50% efficient if half of the energy burned is converted into useful work.[14]

The third component – resulting from the market's feedback loops – has been only sporadically studied. Emanuel Derman, a former physicist working for Goldman Sachs, observed that "in finance, you are playing against people who value assets on the basis of their feelings about the future. These feelings are ephemeral, or at best unstable."[15] I suspect, also, that this third component is a moving target. The amount of feedback we see in the markets has no doubt been profoundly affected by online stock trading, because of the free, high bandwidth availability of prices, analysts' reports, and Internet stock trading sites. Because of these feedback loops, we know that stock prices are not normally distributed, as tends to be assumed in many stock models. In fact, they may be highly kurtotic, or may not have variances at all (though, because it is inconvenient to analysts, this is a fact that is regularly ignored).

STRUCTURE OF THIS BOOK

This book lays out the structure, components, and application of the Financial Dynamics system of valuation. It is designed to provide managers, investors, and other stakeholders with an accurate and comprehensive tool to answer the question, "What is the value of a technology product, project, or firm?" Figure 1.1 shows the presentation of topics in this book. The book is arranged into four sets of three chapters. Chapters 1 through 3 describe the history and characteristics of orthodox measures of "value" – that is, tradable commodities, real estate, gold, and other storable assets. These chapters describe why these valuation orthodoxies arose, and how they have failed to capture most of the value creation going on in the knowledge economy.

Chapters 4 through 6 lay out the components of the Financial Dynamics methodology for valuing knowledge-intensive, technology firms. Chapter 4 describes a mapping for business models in an era in which business has eschewed vertical integration and embraced an ever-increasing array of complex market mechanisms to lubricate supply chains, circumvent capacity constraints, and keep costs low. Chapter 5 explores how ownership and property rights should be approached, given the difficulty of owning "ideas." Ownership has become an especially prickly issue in the wake of the Enron scandal, which effectively hid massive losses by fudging its ownership of so-called special-purpose entities. Chapter 6 then summarizes the components of the Financial Dynamics valuation methodology. Financial Dynamics puts strategic and technology value drivers at the center of value creation and estimation, and is able to effectively calculate valuation risk in the context of these choices.

Chapters 7 through 9 detail the articulation of technology choice and market strategy, and show how to generate value estimators for use in decision-making. Chapter 9 shows how to aggregate this data for managerial, investment, and credit decision-making.

Chapters 10 through 12 explore the applications of Financial Dynamics. Various alternative valuation approaches have boasted of their ability to compute value using publicly available financial data. For example, Boston Consulting Group, Collins Stewart, and HOLT Value Associates have claimed this for their price-level adjusted DCF models; and Stern Stewart, L.E.K., and Marakon Associates have made similar claims for their specific variants of the residual income model. Yet, rather than this being an asset (that is, for anyone but the preparer of the valuation), it is likely to be a detriment to the validity of the model – a fact that became painfully obvious when a number of these firms showed how their approaches "explained" inflated dot-com and telecom values in the late 1990s. The real advantage of having an independent accountant or analyst prepare a valuation is the presumption that they will have access to better data than what is publicly available. Indeed, this is true for auditors (who may or may not reveal it in their reports) and securities analysts (who may only reveal it to their best clients). Chapter 10 shows how Financial Dynamics can be applied under the less than ideal constraints of using only publicly available data. Chapter 11 shows how current audit practice is inherently flawed, and what is needed to effectively address the dictates of the knowledge economy. Finally,

Figure 1.1 Structure of the book

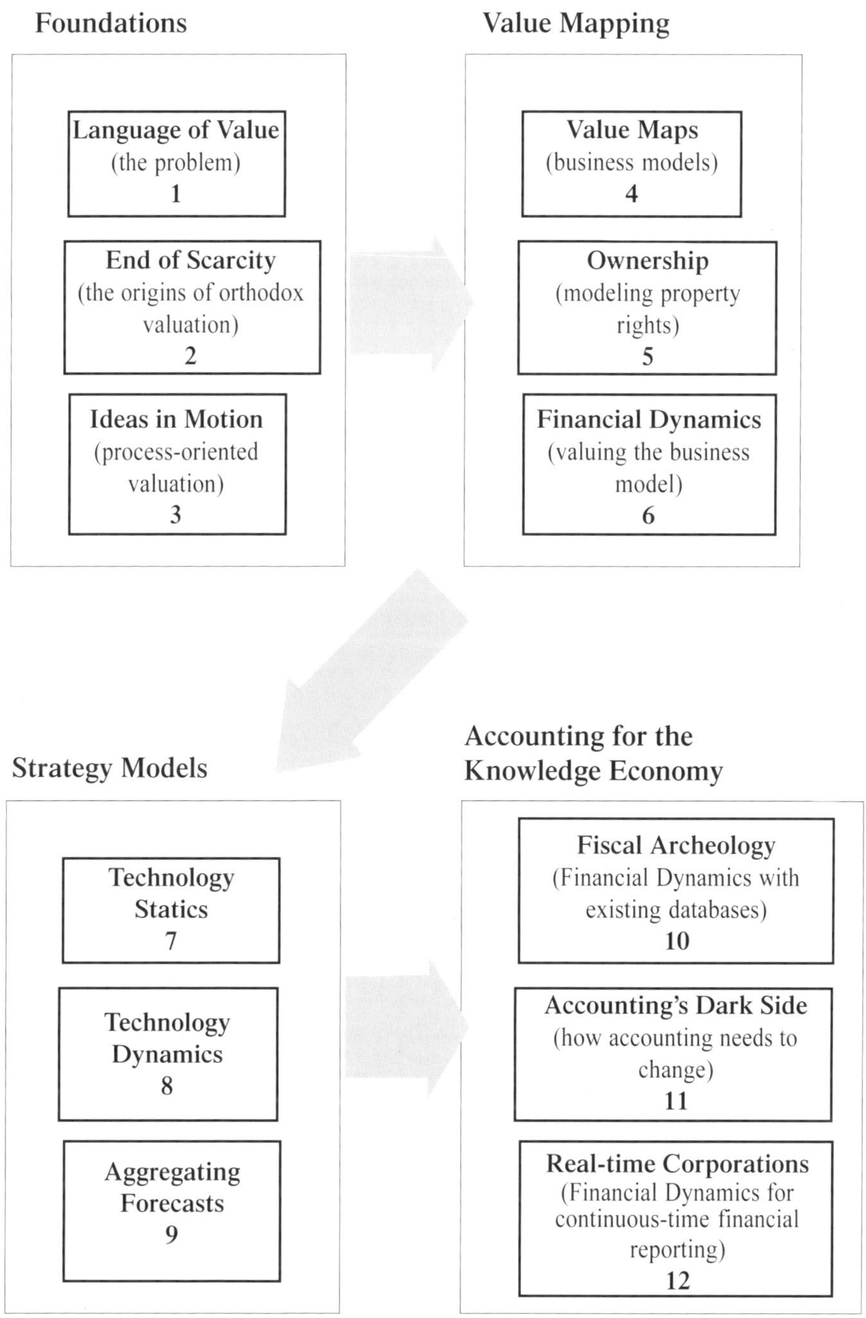

Chapter 12 describes how Financial Dynamics can be implemented as a real-time, continuous accounting system for firms, managers, investors, and creditors that need to make decisions on a daily or hourly basis – not the annual basis of accounting's agricultural heritage.

This book is intended for an audience with some previous exposure to the tools of financial analysis, as well as a basic understanding of the knowledge economy. For those of you looking for background reading in Bayesian statistics, *Optimal Statistical Decisions* by the late Morris DeGroot provides the definitive reference for Bayesian constructs used in this book. *Information Rules*, by Shapiro and Varian, provides an excellent survey of "new economy" concepts and strategy from an economist's perspective; *Valuation*, by Tom Copeland et al., and *Damodaran on Valuation*, by Aswath Damodaran, provide guides for the classical DCF approach to valuation; and for those of you who need brushing up on accounting concepts, try *Accounting the Easy Way*, by Peter Eisen.

APPENDIX: NOTATION USED THROUGHOUT THE BOOK

The mathematical notation I have used in this book derives from the financial economics literature. Unfortunately, this literature spans several disciplines, each with their conflicting notational conventions. Thus, I have, somewhat arbitrarily, used the following notation throughout the book.

Uppercase letters represent the upper limits of the subscripts. Where possible, I use the same letters for upper limits as for the subscripts themselves. Thus, subscript t runs from 1 to T, subscript k runs from 1 to K, and so on. An exception is that I will let subscript i (usually denoting an asset) run from 1 to N because this notation is so common. I use t and r for time subscripts; i for asset subscripts; k, m, and n for lead and lag subscripts; and j as a generic subscript.

I use the timing convention that a variable is dated t if it is known by the end of period t. Thus, g_t denotes a return to an asset held from the end of period t-1 to the end of period t.

I use Pr(•) to denote the probability of an event.

Where the professional literature uses many specialized terms, these are italicized when they first appear in the book.

I indicate differentiation with a subscript – for example, $\frac{\partial Z}{\partial Y} = Z_Y$

Where the variable that the derivative is with respect to is clear, a prime Z', Z'', etc may be used, I indicate differentiation with respect to time, with a dot – for example, $\frac{\partial Z}{\partial t} = \dot{Z}$

Hats (carets) denote sample estimates, so if Z is a variable, then $\hat{Z}$ is its estimate.

Random variables are notated with a tilde $\tilde{Z}$.

---

[1] Quoted by Jeffrey Garten in "Corporate Standards," *Business Week*, May 13, 2002.

[2] Quoted in "Badly in Need of Repair," *The Economist*, May 2, 2002, p. 67.

[3] Economist Jay Ritter estimates that the loss from underpricing of initial public offerings (IPOs) of dot-coms (that is, the "money left on the table") was around US$66 billion in 2000. Often this was recouped by issuing firms through extra commissions (around 5% to 15% of the "money left on the table") or by favored investors through "laddering." For example, in return for being allocated 10,000 shares at an artificially low IPO price, the investor promises to buy additional parcels of shares in the aftermarket at higher prices. By the start of 2002, plaintiffs had filed more than 1,000 lawsuits accusing such malpractice against 40 investment firms.

[4] One of the then-called Big 8 accounting firms; a number that has since dwindled to four.

[5] Quoted in "Special Report: The Trouble with Accounting," *The Economist*, February 9, 2002, p. 61–63.

[6] Quoted in Byrnes N., and Henry D., "Confused about Earnings?," *Business Week*, November 26, 2001, p. 50–54.

[7] The name "guild" comes from the name for a defensive caravan: *Gilde* or *Hansa* in the Germanic countries of Europe; in countries speaking languages derived from Latin, the term was *caritas* or *fraternitas*. The name reflects the early need for physical protection among guild members.

[8] N. Byrnes, D. Henry, and M. McNamee, "Confused About Earnings?," *Business Week*, November 26, 2001 (cover story).

[9] Arthur I. Miller, "Erotica, Aesthetics and Schrödinger's Wave Equation," in G. Farmelo (ed.), *It Must be Beautiful* (London: Granta, 2000), p. 80–101.

[10] A. Aston, "Brainpower on the Balance Sheet," *Business Week*, August 26, 2002, p. 66–67.

[11] Edmund Morris, *Theodore Rex*, Random House, 2001.

[12] Quoted in "A Survey of International Finance," *The Economist*, May 18, 2002, p. 15.

[13] J. Y. Campbell, A. W. Lo, and A. C. MacKinlay, *The Econometrics of Financial Markets* (Princeton, N.J.: 1997, p. 24.

[14] For instance, Charlie Plott at California Institute of Technology and Vernon Smith at the University of Arizona have developed computer market models that look at feedback effects.

[15] Quoted in "A Survey of International Finance," *The Economist*, May 18, 2002, p. 16.

CHAPTER 2

The End of Scarcity

Imagine an underground chamber, like a cave with an entrance open to the daylight and running a long way underground. In this chamber are men who have been prisoners there since they were children, their legs and necks being so fastened that they can only look straight ahead of them and cannot turn their heads. Behind them and above them a fire is burning, and between the fire and the prisoners runs a road, in front of which a curtain-wall has been built, like the screen at puppet shows between the operators and their audience, above which they show their puppets.

Imagine further that there are men carrying all sorts of gear along behind the curtain-wall, including figures of men and animals made of wood and stone and other materials, and that some of these men, as is natural, are talking and, some not. And so they would believe that the shadows of the objects we mentioned were in all respects real.

Suppose one of them were let loose, and suddenly compelled to stand up and turn his head and look and walk towards the fire; if he was told that what he used to see was mere illusion and that he was now nearer reality and seeing more correctly, because he was turned towards objects that were more real, and if on top of that he were compelled to say what each of the passing objects was when it was pointed out to him, don't you think he would be at a loss, and think that what he used to see was more real than the objects now being pointed out to him?

And if he were made to look directly at the light of the fire, it would hurt his eyes and he would turn back and take refuge in the things which he could see, which he would think really far clearer than the things being shown him.

– Plato, *The Republic*
(The Allegory of the Cave)[1]

ACCOUNTANTS OF THE CAVE

Plato's allegory of the cave addresses our perception of what is real versus what is invented. It seems especially apt when describing the world of financial accounting. Six centuries ago, accounting settled on a description of economic phenomena which seemed "real" in a world dominated by agriculture and marine trade in the products of agriculture, such a spices and tea. Over the centuries, industrial production, and then computers, dramatically changed this simple world. The modern economics of commerce evolved and grew complex, while accounting remained trapped inside a cave of its own design, knowing the world outside only through the shadows of its own invention. These latter-day cavemen believe that they can view the real world more accurately by stirring the embers of their fire, or by plastering the wall of their cave.

They are wrong. Consider a bank statement which lists each deposit as a single sum. A deposit of several checks and some cash appears as a single number, making it impossible to distinguish a single check from among them – the information was obliterated in the representation. What's more, the form of a representation determines what can easily be inferred from it, because the symbols and their arrangement are the only things an analyst or other user of financial information can draw inferences from. Our representations of numbers and sums were valuable in the days before computers because addition can be performed on the numbers with a few dronelike operations (a fact which has led to many unfair characterizations of accountants and their personality). But with computers, a vast new array of options for inference has opened up; if only we knew how to tap into them (the motivation for this book!).

And what are the shadows around which accounting dances? Accounting for firm operations has traditionally involved first putting a price on each "thing" involved in an operation, and then summing these numbers. The more accurate approach would keep the qualitative information associated with raw materials, assemblies, labor, finished product, services, and so forth (that is, the hedonics) and assign values to them only at the end of the accounting process (that is, closing). Or better yet, allow the users of financial statements (which could conceivably be in the form of software and computer data) to flexibly assign their own values. The latter approach involves significantly more complex calculations, but this becomes more or less irrelevant when computers are available. An Excel spreadsheet would be adequate to present firm data. Presenting the qualitative characteristics of value flows in the firm makes it much more difficult for analysts to play games with what is happening in customer and vendor product markets.

You might be justified in asking why I mention Plato at all? Because, until late in the Renaissance, three Greek philosophers – Socrates, Plato, and Aristotle – defined much of the received wisdom of the world, including business, economics, and accounting. The accounting methods of the scholar and friar Pacioli were more or less set long before scholars dared to budge from Aristotle's teachings. The rule-laden perspective of Aristotle very much dominated Pacioli's old science of wealth – double-entry accounting.

But of the Greek triumvirate of philosophers, my choice is Plato, whose ideas and demeanor seem most attuned to the world of the 21st century. Indulge me in a bit of history. The first of the three, Socrates, was the son of a sculptor and a midwife. He refused to write down his ideas, insisting that to do so dulled the memory. He claimed that knowledge would reveal itself through questions and answers – what he considered a form of intellectual midwifery. His pupil, Plato, was a champion of reasoning through problems in a constantly shifting world, dividing the world into ideas or objectives – which were permanent – and into phenomena, which were measurements or perceptions of those ideas and objectives. According to Plato, you were limited in the information about the world of phenomena that could be gained through measurement and the senses. Only reasoning through a problem or situation goes straight to the idea. Plato had no qualms about writing his ideas down – indeed, all we know of Socrates comes from Plato's dialogues. Plato recognized the limits of human memory, and recorded his insights on parchment "databases" in order to pass them on to future generations. Plato's star pupil, Aristotle, carried on this tradition, though he took fundamental issue with Plato's philosophy. The son of a physician, young Aristotle followed his father in making detailed taxonomies of the anatomies of animals and their behavior in the wild. These taxonomies engendered a "bookkeeping" mentality, and a dedication to rigid rules of logic that distinguished him from the more inquisitive Plato. Aristotle's rule-laden taxonomies segued with the designs of the medieval church and reigned well into the modern era.

Plato encouraged his students to "see" his ideals and objectives; then to argue with him about the processes and phenomena that one must encounter to reach them. Socrates never allowed himself to be pinned down, leaving little of permanence to guide posterity, and ensuring his imprisonment and death after humiliating one too many Athenians. Aristotle's method encouraged the mindless pursuit of detail, with little regard for insight or evidence. It took Galileo's invention of the empirical method to reverse the influence of Aristotle. For his discoveries, Galileo was excommunicated – a verdict that the Church overturned only at the end of the 20th century.

Needless to say, in this book we embrace the Platonic method – argument and discovery centered on ideals and objectives. It avoids the evasion and weasel words of Socratic method. It eschews the mindless checklists and structure of Aristotle. It seeks a holistic picture of the business, and valuation that satisfactorily meets the needs of stakeholders in the firm.

Once we have defined our business objectives, then the best insurance of success in future decision-making – which is what accounting is supposed to help us do – is to build hypotheses based on our measurements and uncertain understanding of the phenomena around us. Aristotle's vision of a structured and rule-laden world simply does not mesh with the manager's world, where uncertainty, short deadlines, and fugitive information prevail.

HOW WOULD "VALUE" CHANGE IF YOU HAD A NANOBOX?

Ichiro Asai is one of the best motorcyclists in the world, having won the 2001 SuperSports 600cc class FIM Asia Road Racing Championship. He also has one of the most daunting "day jobs" in the world – as a test rider for Yamaha Motorcycle's prototypes at the factory's Sugo racetrack near Sendai, Japan. The softspoken Asai described his job one afternoon during some free time at the Sugo track. "The bikes need to be driven at 110% of their capabilities. We really push the bikes past their limits. The test driver's job is to discover design flaws, whether it is in the chassis, motor, suspension, ergonomics, or any of a thousand other aspects of the bike." None of this seems particularly surprising until Asai relates the terms of his contract with Yamaha. "I have my job up until the first time that I crash the motorcycle. Any damage to their prototype and I will be fired immediately." This is a bracing revelation in light of the fact that motorcycles, unlike most other forms of transportation, suffer what engineers call "capsize instability" – they fall over.

Yamaha spends millions of dollars on R&D, prototyping, and homologation which can take up to 10 years and tens of thousands of test laps in a typical design. Their prototype is an extremely expensive "design database"; one that is not easily replicated. Once the design is stable, production is simply a matter of stamping out the finished motorcycles. In motorcycles and many other tangible goods, the majority of the cost is incurred up front, in the generation of ideas – not at the end, when materials are being used. In fact, the cost of materials continues to decline, and is around half the material of a typical 1970s-vintage motorcycle.

The economics of motorcycles today have become – for all intents – just like those that apply to software and other intellectual products. And the list of products that have shifted from the physical realm to the realm of ideas continues to grow as computer circuits and software substitute for the traditional gears and cams of watches, cameras, automobiles, and other household products. Technology for printing circuits with newsprint technology, and creating products out of metal power from a database of design templates that have already been prototyped, and will soon be in production. The age of the nanobox is near. The conundrum of modern valuation is the trend toward unstorable wealth – wealth is becoming solely a function of your future ability to generate value.

A variety of well-funded projects today look forward to the day when matter will become software. As a result, we will be able to use the Internet to download not just software but hardware as well. Nanotechnology – research into machines smaller than a few billionths of a meter – moves automated assembly down to such scales that manufacturing evolves from a physical process of combining capital, labor, and materials, to an intellectual process of design and programming on a nano-scale. The first applications have begun in disk drives and semiconductors; these will spread to small consumer products and other knowledge-intensive products.

The ultimate goal of matter-as-software research will be the nanobox – sort of a futuristic copy machine that combines nanotech fabrication with flexible desktop-manufacturing methods used today for quick prototypes of new products. You could purchase a recipe on the Internet, and then insert a sheet of plastic and squirt electrically conductive molecules into the "toner" cartridge. The nanobox would pass the plastic back and forth, laying down patterns of molecules, then electrically direct them to assemble themselves into circuits. Next, using different "toners," the nanobox might add other components and finally build up a housing.

If traditional accounting models have difficulty with the knowledge-intensive businesses of today, they can be expected to fail completely in an economy of information industries. In fact, were the nanobox available today, conventional financial accounting would very likely become useless overnight. In practice, as portions of the technology become available, we will find portions of traditional accounting growing irrelevant – or worse, misleading. Indeed, this has been happening for over two decades.

The first nanobox is already being deployed by the Pentagon; called the laser engineered net-shaping (LENS) tool, it is designed for battlefield use. If a vehicle, for example, breaks a component, a technician can search a database for digital blueprints, either on board or by satellite-link from a U.S. maintenance center. The LENS deposits a fine path of powdered metal, tracing out cross-sections of the part-in-progress. Trailing behind is a computer-controlled laser that fuses and hardens the metal particles. Layer by layer, more material is fused into place, so that – at a rate of about one cubic inch of material per hour – a new part emerges from the smoke and dust. The technology is expensive and slow, but is justified in the battlefield where there are significant consequences of not having a replacement part.[2]

With the prospect of "virtual" production from nanoboxes on the horizon, accounting's traditional valuation based on inputs – materials, labor, capital – needs to be rethought. The Army's LENS nanobox carries its field inventories in its database. It obeys the same economics as a software program – the first copy is expensive, all additional copies are free or almost free. Surely this demands a different definition of value.

THINKING OUTSIDE THE NANOBOX

Our orthodox traditions teach us that tangible things are the basis of wealth – only things you can hold in your hand, build a house on, or deposit in a bank have "value." Such concepts of value are relatively recent, dating back only a few millennia. Before there were markets, barter, or money, there was a lively trade in ideas. Richard Dawkins, Professor of the Public Understanding of Science at Oxford University, coined the word "meme" (pronounced like "cream" or "dream") for ideas and other elements of culture that tend to make copies of themselves and are self-replicating, like genes. They are stored in human brains and passed on by imitation. As examples he suggests "tunes, ideas, catch-phrases, clothes fashions, ways of making pots or of building arches."[3] American philosopher Daniel Dennett sees memes as central to evolution and the development of modern culture and commerce. He suggests that economic systems, along with the rest of culture and human consciousness, are a huge complex of memes. Among his well-known quotes are: "... a human mind is itself an artefact created when memes restructure a human brain in order to make it a better habitat for memes"[4] and "A scholar is just a library's way of making another library."[5]

A widely accepted economic definition of the value of a good, service, or other entity is the discounted future value flows generated by them. Very few knowledge economy companies (for example, think of Microsoft or Qualcomm) count their wealth in terms of inventories of stored assets. When goods were mainly fungible commodities that could be stored, this value coincided with accounting's summation of material and labor costs.

Financial valuation based on value flows discounted back to the present reflects consumers' "impatience" – their preference for value now versus in the future. But future earnings are neither reported in financial statements nor freely available in the market for financial information. They have to be constructed through a forecast – a hypothesis, conjecture, or speculative view about something that will occur in the future. Forecasting is a hybrid of science and artistry and is conditional on the forecasters' extrapolations of past experience in light of future strategy.

Such a definition of wealth requires predictions of two classes of future values:

- *Future value flows:* Future value flows are approximations that are functions of prior observed value flows, combined with subjective information about uncertain future strategies, opportunities, options, market forces, and so forth. Future values are never objective, but they can be constructed so that they have desirable and consistent properties.
- *A discount rate:* The discount rate is a measure of the time value of money. It may either be thought of as an *intrinsic psychological*

value – that is, "impatience" – or as an *opportunity cost* – that is, the returns from possible alternative investments. The latter definition is commonly applied and requires prediction of uncertain future opportunities. Future discount rates are never objective; thus, it may be desirable to compute value based on a schedule of possible discount rates.

A BRIEF HISTORY OF ACCOUNTING

The noble savage enjoyed a brutal certainty long since lost in the accounting of his modern progeny – the certainty of the club and the spear. Property negotiations were conducted messily, but with the virtue of complete transparency. Settlement prices – the number of severed limbs and cracked skulls – yielded to quick tally. Accountants, analysts, forecasters, and the like might have been useful as sources of supplementary winter protein, but not much else.

As man came to depend increasingly on factors beyond ken and village gate, a policy-making elite were called on to render meaning in a complex world. In fact, the hording of knowledge was one of the original paths to power and wealth. The priest-cast that first cracked the seasonal Nile flooding used their information to overawe subjects with their "magic," leaving farmers hanging on each pronouncement before planting. Early commerce reinforced a tributary system, lubricated by "mathematical magic" whose sole purpose was to enrich and entrench the priests.

The first great commercial markets arose in the Middle East, in Sumer, from around 3500 BC, initially as part of a barter economy. Sumerian cuneiform script developed around primitive accounting that accompanied their ancient markets. Scribes were professionals situated at the entrance to a market, who would update a history of trades between individuals, written in duplicate on clay tablets with reed styluses, and "signed" by both traders with the seals worn around their necks. Written language before the Greeks was almost exclusively the domain of accountants, and generally lacked the descriptive breadth required for literature. There was no mathematical system for accounting in this age, as few people could count or add, and mathematics was consigned to the realm of magic and astrology. Accounting was indistinguishable from writing, and accounting statements were simply rambling inventories of meetings and trades. Indeed, the Sumerian goddess of accounting, Nissaba, was also the goddess of writing and scribal arts.

This balance of power shifted dramatically in the first millennium BC, after the Lydians standardized the weight and gold content of coins. Standard weight was assured by Croesus's treasury which weighed the spherical nuggets, then stamped them with the treasury seal, which forced them into the flat, round shape of coins that we retain to this day; purity was "audited" using the black touchstone – rubbing gold against the touchstone left a specific discoloration which was distinguishable from the discoloration left by gold mixed with less valuable metals. Once standard coins existed, merchants, rather than depending on personal judgment when bartering one good for another, could price their goods by "tale" – tallying up the individual worth of goods in these standard measures – thus lifting mathematics out of the realm of magic and priests, and democratizing it firmly in the arena of commerce. This process was rendered in Latin as *ad- + cunter*, to count (initially from *computare*, to sum up), which is the root for our modern "accounting."

Coins provided two inventions essential to accounting: (1) a *standard unit of measure* for value by which all commodities could be valued – that is, the weight of a coin; and (2) the introduction of *counting and mathematics* into accounting. The mathematics involved in summing up the value of possessions became a national pastime in Lydia, leading to many tales of the wealth of Croesus. *The Lydian objective for accounting was to compute totals of wealth.*

The growth and complexity of commerce engendered systems of accounting. Accounting systems were often intended to provide information that managers and third parties could use in decision-making. They also served other purposes: to produce operating documents; to protect the company's assets; to provide data for company tax returns; and, in some cases, to provide the basis for reimbursement of costs by clients or customers. The nature and variety of objectives has changed significantly over history.

Greece and Rome added little to the methodology of accounting; their clumsy number systems actually impeded mathematics. It was not until the end of the first millennium, with the ascent of Sylvester II to the papacy in 999, that accounting saw innovation. Sylvester II was held in high repute for his learning, especially in mathematics and astronomy, and revolutionized medieval accounting by introducing the use of the abacus, Arabic numerals, precise time and date reckoning via astrolabes and (legend has it) pendulum clocks. His treatises on the astrolabe and the abacus introduced multiplication, division, percentages, and other modern business procedures that we now take for granted. Unfortunately, his scholarship did little to endear him to his flock; rumors circulated that his knowledge of the methods of commerce and mathematics was proof that he was a magician in league with the devil.

Sylvester II's reign ended after three years, and the introduction of modern accounting, arithmetic, and timekeeping were delayed for several centuries until the 14th-century monk Pacioli codified these and other accounting advancements in a widely read book. From this sole purpose arose many contemporary conventions: historical costs, arm's-length transactions, and debit and credit (meaning "debt" and "entrust," respectively). Yet, ultimately, *Pacioli's "double-entry accounting" served only one objective: to compute the wealth of traders*.

Pacioli's book made his methods of error control and principles of valuation available to a wide audience, and this influenced the evolution of accounting objectives. In medieval Europe, accounting demanded considerable labor, performed by scribes in an era when few could read and write, and where those who could felt commercial pursuits beneath them. As the need for information of this sort grew widespread, accounting broadened its objectives, while economizing on the recording and mathematics. *English manorial accounting in the 16th and 17th centuries promoted "managerial" objectives for accounting, in addition to "wealth tracking."*

Accounting objectives evolved considerably in the 19th century with the growth of railroads – the great network of that day. To manage the vast and dispersed inventories of tracks, ties, cars, and engines, accountants adopted a dizzying array of "surpluses" and "encumbrances" (terms now obsolete) which were the forerunners of modern *accrual accounting*. Accrual accounting didn't just track inventories of possessions; it also looked at potential future cash outlays for the replacement of physical assets, such as ties and track, as they wore out. Surpluses and encumbrances presaged modern accrual concepts of depreciation, allowance for uncollectibles, and so forth. *Wealth tracking evolved from simple inventories to complex estimations of future firm activity*.

The latter-day version of the wealth-tracking function is what is called the stewardship function of accounting. This function dominates accounting objectives as promulgated by the Financial Accounting Standards Board.[7] Stewardship accounting focuses on providing information that enables investors and creditors to evaluate how the firm has performed.

Two more recent accounting fields, product costing (cost accounting) and managerial accounting, emerged in the middle to late 19th century. Cost accounting provided a tool for production engineers (early articles on product costing appeared in mechanical engineering journals). Managerial accounting, in which more attention is paid to economic decision-making and the economic analysis of costs, is even more of a recent addition, developed in the 1960s, with a focus largely on budgeting and cost control. Both were designed to provide better valuations of the business processes as technology and business processes grew more complex.

A fourth variety of accounting – tax accounting – identifies "income" as a basis for government collection of tax revenues. Tax accounting serves three widely divergent objectives: (1) it provides data input for computing national accounts; (2) it enacts social policy, through subsidies and taxes on certain behavior, such as pollution or marriage; and (3) it collects revenue to finance running the government.

The *Securities Exchange Act* of 1934 selected one of these many varieties of accounting, stewardship, as a first among equals – what we have come to know today as GAAP or financial accounting – giving it legal credence. Though acceptable for the industrial economy of the 1930s, its shortcomings grew over time, and banks and investors demanded additional information in the form of credit analyses, managerial and cost accounting, and securities and business valuation. In each case, substantial additional fundamental information had to be gathered. (Securities analysts typically have preferential access to a variety of internal information in the firms they cover, but a firm usually extracts a price for this access: that reports about the company be bullish.) The fragmentation has hampered our ability to effectively value businesses, because it splits what should be unified accounting and reporting into a system of mutually incompatible and uncooperative fiefdoms – isolated "stovepipes" which impede efficient and effective financial decision-making. No wonder so many different interpretations of firm value are possible, and even coexist at any one time.

My cursory history of accounting illustrates the varied and evolving set of functions that over time have been supported by accounting. Despite the dramatically divergent objectives of different flavors of accounting – and these are growing even more varied, complex, and divergent in the knowledge economy – the raw data for each set of corporate accounts derive from a common system for data collection, error control, and summarization. Too often, they are also drawn from a common methodology for account calculation, which can severely compromise the informativeness of accounting. Financial Dynamics seeks to rectify these shortcomings with a system that can efficiently support a broad range of objectives from common data sources.

FROM SCARCITY TO ABUNDANCE

Our concepts of wealth, income, and value have been for the most part molded by the economics of scarcity – scarcity of manpower, machines, materials, or money. In contrast, the knowledge economy is shaping our future on an economics of abundance – abundance of technology, knowledge, tangible assets, acquaintances, attention, and leisure. This will only accelerate as information technology and robotics continue to streamline our workplace and usurp traditional jobs.

The knowledge economy requires a new science for valuing wealth – one in which the ongoing activities of the firm, and the demand-side market forces, dictate the accounting effort. The traditional supply side, historical cost accounting for scarce resources – materials, manpower, and machines – can no longer provide the information needed for management to make educated choices. Our new science must embrace a measure of the scarce resources of a knowledge-intensive business – attention, intellectual property, and knowledge capital.

How would such a science of wealth look? The various academic studies of finance and accounting form, collectively, a branch of the social sciences. As such, there are specific strains and biases to formal research in these areas. In particular, social science research tends to value discovery and prediction via objective survey of the real world. This is in distinct contrast to the physical sciences and engineering, where discovery and prediction take a back-seat to modeling and methods. Technology pundit George Gilder once observed that "like the Pentagon, social science reduces all phenomena to dollars and body counts."[7] Nothing wrong with that; but such a perspective has its limits. Economist Friedrich von Hayek observed that "there are no better terms available to describe [the] difference between the approach of the natural and the social sciences than to call the former 'objective' and the latter 'subjective.' The reason for this is that the object, the 'facts' of the social sciences are also opinions."[8] The fact that finance and accounting are inherently "subjective" – a consequence of their social science pedigree – creates significant problems for disciplines which, by design, are supposed to be objective.

We may have little choice in this, but objectivity is also a matter of degree. The sciences, along with economics and mathematics, have evolved significantly since the medieval origins of accounting, giving us new tools with which to improve the objectivity of accounting. But social sciences are not only saddled with opinions of fact; they are also saddled with religious, political, and even gratuitous opinions. Accounting must, additionally, deal with the dictates of property law – a particularly contentious social science in the age of intellectual property. Accounting's referent discipline – economics – is dubbed the "dismal science" as much for these shortcomings as for its focus on "dollars and body counts." In my previous book, *Valuing Technology*, I surveyed the technologies that would continue the current drive toward knowledge-based wealth. More recently, *The Economist* polled its readers, asking what technologies they thought would propel the next Industrial Revolution (Fig. 2.1).[10] Their poll revealed few surprises (at least for your author). Most respondents bet that information technology would dominate – if for no other reason than other technologies are so knowledge-intensive that computers will be essential in exploiting them. Biotechnology was even forecast to evolve into a branch of information technology. Though physicists continue to argue about what quantum theory really means, the rest of us have no qualms about working with it. Perhaps 30% of the US GDP derives from technologies based on quantum theory.[11] Without the insights of quantum mechanics, there would be no cell phones, no CD players, no portable computers. Readers believed that computers would outstrip the human intellect within the decade – an event horizon known as the Kurzweil Singularity[11] and discussed in *Valuing Technology*. Keeping such hyper-intelligent machines in check was a concern for many respondents. In contrast, the technologies that have most affected product and labor markets over the past decade – those of geographical and organizational scaling – were seen as less important, reflecting the perceived decline in economic importance of physical assets and manual labor. One technology I mentioned in *Valuing Technology* which, oddly, seems not to have registered with the public was that of fractals – Benoit Mandelbrot's invention describing self-similar, self-replicating processes. The field of cellular automata studies similar models, and has recently received a stimulus with the release of Stephan Wolfram's A New Kind of Science. Wolfram's work describes the manner in which simple, recursive models can effectively describe many of the complex phenomena of nature. These models hold particular promise in the social sciences, where mathematics designed primarily to serve the natural sciences have consistently failed to show acceptable predictive or descriptive power.

Figure 2.1 Technologies of the sixth Industrial Revolution

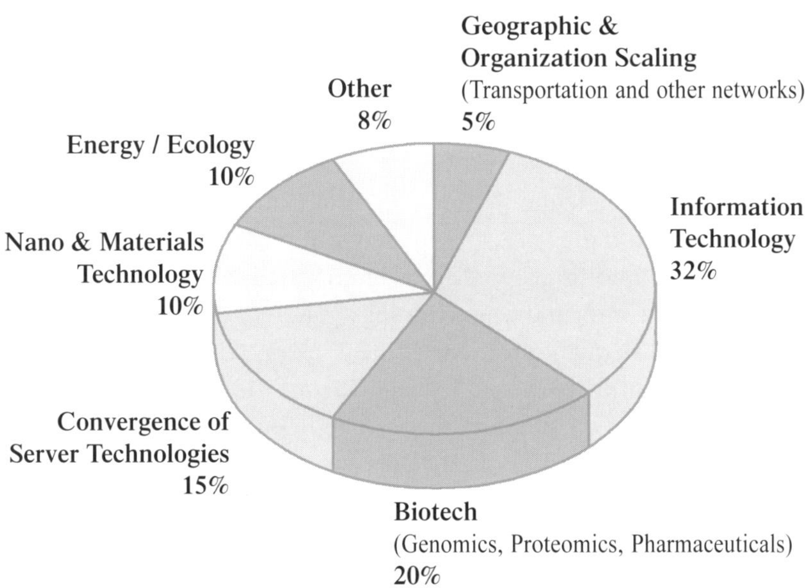

The advances in technology businesses have undermined GAAP's "stored wealth" approach to valuation. The breakdown in GAAP's "stored wealth" approach to valuation is illustrated most dramatically in the divergence of the accounting profession's valuations and the price that people are actually willing to pay for a firm. For example, the book value of the S&P 500 firms is now less than one-fifth the value that investors pay for them (Fig. 2.2). Oddly, some accountants have suggested that accounting values cannot be compared to stock-market values because the latter are somehow not "real" or "objective." This explanation seems peculiar in light of the fact that stock-market valuations reflect an amount of hard cash relinquished for ownership in a company. You can hardly get more "real" than that.

Figure 2.2 Accountants track less than one-fifth of the firm's value

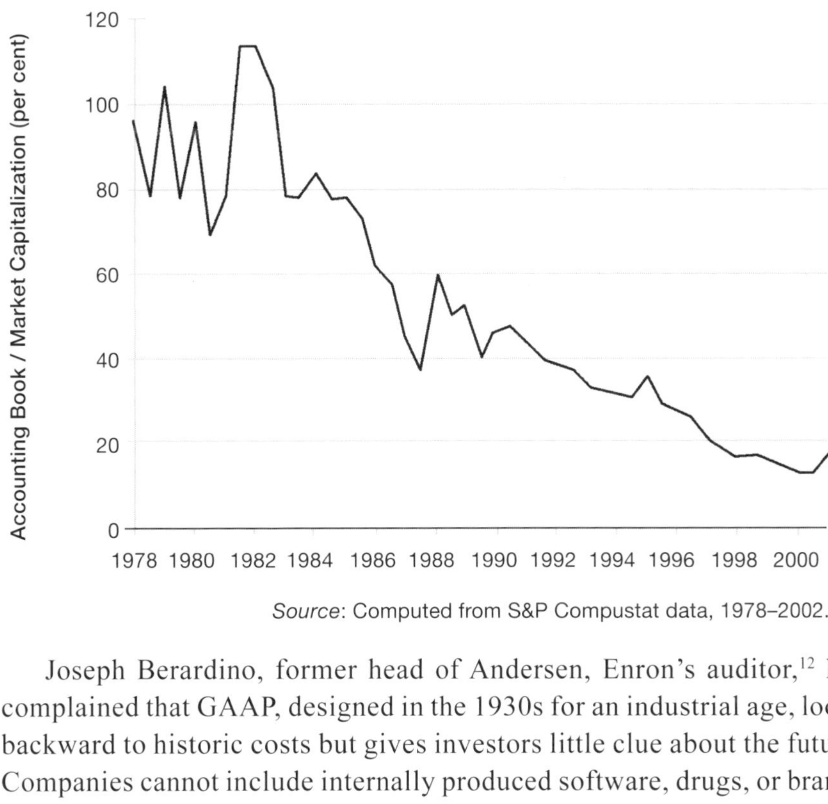

Source: Computed from S&P Compustat data, 1978–2002.

Joseph Berardino, former head of Andersen, Enron's auditor,[12] has complained that GAAP, designed in the 1930s for an industrial age, looks backward to historic costs but gives investors little clue about the future. Companies cannot include internally produced software, drugs, or brands in their balance sheets because they are intangible assets. That has led to an increasing gap between the value of companies as measured by the stock market and the value measured by their accounts (their book value). Economists use the price-to-book ratio as a close indicator of Tobin's Q, the ratio of market capitalization to the replacement value of the underlying corporate assets, which is difficult to compute in practice.

Corporate valuation serves no single objective, but in the ideal needs to effectively address the conflicting requirements of many constituencies. Nor can we expect valid models to unequivocally specify a single "true" value for the firm. This is asking too much of our models and is inconsistent with the modern perspective adopted by the accounting profession. The accounting profession abandoned the goal of expressing "true" income in the 1950s, adopting instead a "stewardship" objective which to this day is promulgated by the FASB. This perspective focuses on providing information that enables investors and creditors to evaluate how the firm has performed. Unfortunately, in recent years, the accounting profession has failed broadly in this regard, and there is now little relation between accounting value and market value – that is, the amount that people are willing to pay out-of-pocket.

The "stored wealth" and "value flow" perspectives of corporate value modeling have their precedents in accounting theory. Accountants have classically divided their chart of accounts into real and nominal accounts. The revenue and expense accounts comprising the income statement are called real accounts – real in the sense that they capture the activities in the operation of the firm. The remaining accounts – that is, the nominal accounts – relate to the firm's stores of wealth, and the various claims against those stores.

The two are related through concepts of ownership and cost. A cost is a cash outlay or other sacrifice of resources. Costs which are expected to generate future revenues are called assets – that is, future stores of wealth; otherwise they are expenses – that is, value outflows in the current accounting period. Liabilities are claims by creditors against the assets; equity represents residual claims by the firm's owners. Obviously, the dichotomization of costs into assets and expenses is highly subjective, since it involves estimating an uncertain future. Accountants have given up any claim to accuracy, instead blindly invoking time-honored depreciation methods (usually linear).

Ownership is dichotomized into owner's equity (firm wealth owned by shareholders) and liabilities (claims against firm wealth by outsiders). The total store of wealth is increased or decreased by the revenues net of expenses, also called income.

This traditional perspective reveals a process orientation even to traditional accounting. It also places emphasis on the need for forecasting future revenues, which is required for dividing costs into expenses (that is, expired costs) and assets (that is, stored wealth).

The preference of GAAP accounting for the stored wealth perspective has arisen more for reasons of convenience and efficiency than for validity and accuracy. Before the days of computers, the reconciliation of accounts and search for errors was too time-consuming to allow its application to all transactions that flowed through the firm. Instead, a macro-error control was the original justification for double-entry accounting. Thus, a shortcut was devised. At periodic points in time (for example, annual close, physical inventory), corporate stores of particular types of wealth (for example, the nominal accounts for inventory, accounts receivable, cash) would be counted, and compared to the residual amounts computed on the general ledger. They would be adjusted until auditors were convinced that accounting for the real accounts was valid. The technique was otherwise wasteful, labor-intensive, and time-consuming.

Advances in computing over the past two decades have made possible the real-time audit of value flows, thus obviating the need for periodic counts to validate nominal accounts. Advances come in a variety of forms – perpetual inventory systems, real-time credit tracking, and so forth. Yet, external audits still are conducted through counts of stored wealth, leaving many opportunities for error in reporting.[13]

DO ANALYSTS UNDERSTAND VALUE BETTER THAN ACCOUNTANTS?

One of the main reasons for the divergence in how securities analysts value companies concerns perspective. Most analysts' formulas use *future* earnings or cash flows as the basis for valuation – discounting them back to the present. In the 1990s, this led to wild assertions about the future potential of Internet business. Accountants looked in the other direction – their historical cost objectivity limited them to a myopic (in the minds of the securities analysts) and backward-looking (the accountants called it "historical") perspective. The result was a theatre of the absurd – securities analysts assigning fantastical valuations to companies that accounting statements claimed had negative net worth.

How good are analysts at using existing accounting statements, time series of cash flows, and earnings streams to predict future earnings? Modern valuations of companies always reflect some time discounting of future earnings. Since any forecast is subject to legitimate forecast error – not to mention moral hazard and other incentives to "fudge" – ability to value a company very much depends on how good is our "language of value." Let's first look at forecasts the experts produce. Figure 2.3 tracks the deterioration in earnings forecast accuracy over the past decade for consensus forecasts of the S&P 500 tech stocks.

Figure 2.3 Consensus of annual earnings forecasts for the S&P 500

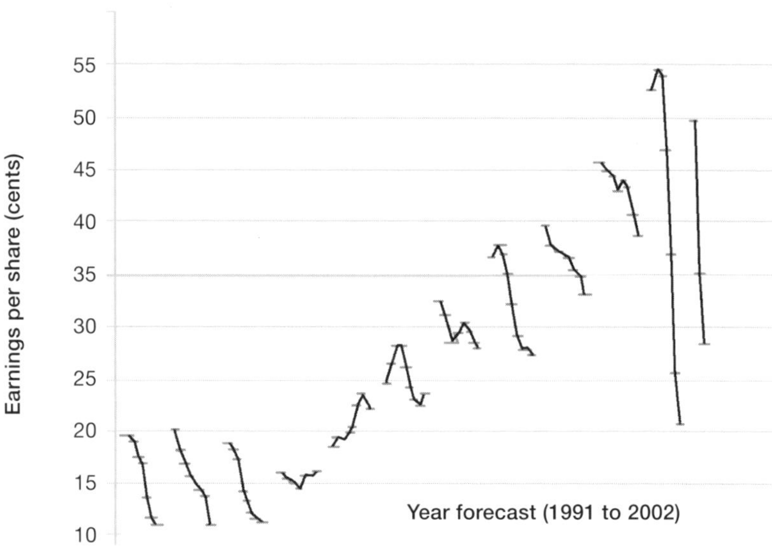

Source: Yardeni.com.

This estimation task, restricted to a very short time period spanning no more than two years at maximum and unencumbered with cost of equity capital or discounting, must be a walk in the park compared to forecasts 10 years out (a normal staple of the DCF calculations). Yet, within this trivial two-year period of time, a 100% revision of forecasts is quite normal by the best forecasters in the country, and these forecasters experienced even greater errors in the uncertain post-dot-com climate. The widespread use of consensus numbers has tended to stop analysts coming up with low-ball numbers for fear of facing the wrath of investment banking colleagues chasing deals from the companies they are writing about. It has even prevented them from pointing out the chasm between pro forma and GAAP. This is perhaps the reason that, historically, "sell" ratings have constituted fewer than 1% of analysts' recommendations, according to Thomson Financial/First Call.

Error is not the only problem; the potential for moral hazard is obvious. Analysts generate a veritable stream of recommendations – buy, accumulate, hold, and so forth – and are seldom independent. Either they have their own inventories of securities, which they profit from; or they have been co-opted by the public relations arms of the firms that they are rating. An analyst who releases a "sell" recommendation is certain to be cut off from a direct flow of information from the corporation – suicidal in the business of financial analysis. So seldom is a "sell" recommendation made that it is typically industry news and a "hold" is considered the equivalent of a "sell." What analysts publish, and what they tell their richest customers over the telephone, are often divergent.

Things change less than we often believe. The business of forecasting has a long and dodgy history. In the ancient world, sorcery came in four forms: (1) divination; (2) sympathetic magic based on imitation, or by making use of associated objects; (3) thaumaturgy, or wonder-working; and (4) incantation, or the chanting of spells, verses, or formulas.

Divination was traditionally the most sought after (and profitable) variety of sorcery. Its greatest ancient incarnation was at the center of the Greek world in Delphi, where ancient Greeks erected the Temple of Apollo to enthrone their own Delphic Oracle over a natural vent of noxious and inebriating gases. For several centuries, this drunken oracle concocted forecasts that were widely sought after in both private and public decisions.

Indeed, the Delphic Oracle delivered probably the first ever "financial" advice. King Croesus – king of the Lydians, the inventers of gold coinage, and the first people to establish permanent retail shops – was encouraged by the oracle to use his wealth to invest in an army of war against the

Persians, predicting "a great kingdom would fall." War was waged, King Croesus lost, and Lydia fell.

Despite their oracle's shabby reputation and dismal track record, the "Delphi" name seems to have gained a cachet of respectability in contemporary business. The Delphi Method is widely promoted for technology forecasting,[14] and Delphi is a popular name for all sorts of intellectual property. There is no suggestion of substance abuse involved in either application.

IS GAAP RESPONSIBLE FOR POOR FORECASTS AND MISLEADING VALUATIONS?

Analysts relying on financial data for forecasts are typically constrained to small data samples – for example, time series of 10 to 20 data points are not unusual. Unfortunately, many of the distributions, models, and methods of statistics assume large samples. This is because these techniques were developed originally in fields such as physics and agriculture, where any limitations on sample size could easily be lifted by running more experiments or growing more crops. Not so easy in business. Once the year is up, managers do not have the option of redoing the year – even with the altruistic goal of rectifying mistakes or gathering additional data.

Another problem with accounting information arises from a much more ancient source. Double-entry bookkeeping was introduced to accounting in the 12th century, and codified by Pacioli in the 14th as a technique to control bookkeeping errors in a time before computers. As its name suggests, every entry is made at least twice (as a debit and a credit). But most accounting transactions appear in different form over the transaction cycle. For example, a purchase order may initiate a vendor shipment and invoice, a trucking transfer and invoice, receiving documents and accounts payable, cash disbursements and adjustments, and returns resulting in debit and credit memos. Controls are implemented to ensure that each of these components of the transaction cycle has exactly the same information as the other, so that they can be recognized as part of the same transaction. Unfortunately for analysts, this duplication of information shows up as multicolinearity in the data. Multicolinearity is a statistician's term for a particular data weakness where all of the data values tend to move in synchronicity, and each additional data point provides only a fraction of the information that one would expect. If accounting research provides any indication of the magnitude of the multicolinearity problem in accounting data, it is the fact that cross-sectional econometric studies in accounting typically exhibit R^2 statistics in the range of 5% to 10%, indicating that accounting data provides only 1/10th to 1/20th the explanatory power of an ideally informative data set.

Multicolinearity and small sample limitations of quantitative financial data demand added emphasis on the qualitative information that is available for corporate valuation. Qualitative information helps define both the structure of the business model, as well as its initial parameter estimates.

WHERE GAAP IS MUTE

A host of accounting techniques – all quite legal, all embraced by GAAP, and yet not presenting a valid picture of their finances – abetted the Enron collapse and severely impaired public confidence in the relevance of accounting to a 21st-century business.[15] Harvey Pitt, former chairman of the SEC (and a former lawyer for the accounting firms Ernst & Young and PricewaterhouseCoopers), said it best: "the current system of disclosure is designed to avoid liability, not to inform anybody."[16] Accountants' monomania with "consistency" and "objectivity" (important for their self-defense in the courtroom) has been made at the expense of "informativeness" (the key reason for their unique and pivotal position). GAAP makes it hard to know exactly what a company earns and thus what its stock is actually worth.

It is business operations over which management has the most control. It is also the only source of risk and uncertainty that accounting statements directly deal with. Where markets are stable, and the firm is competing effectively in the market with products that are accepted by consumers, then any changes in firm value can be assessed by looking at management activities in this realm. Unfortunately, ever since the 1970s, spot markets, outsourcing, rapid coordination of diverse producers, and so forth, have increased the proportion of risk contributed by product and capital markets, making accountants' financial statements less useful in risk management.

Accountants increasingly tend to overstate earnings, drawing into question their independence, and even suggesting moral hazard. Accountants have lost their role as impartial "value assessors" of the firm; that role is increasingly being taken by analysts' groups such as First Call and Standard & Poor's, or consultants promoting alternative valuation methodologies such as economic value added (EVA). This is graphically displayed in Figure 2.4, showing how GAAP consistently overstates earnings when compared to other respected reporting institutions.

Figure 2.4 The divergence of income assessed by First Call, S&P, and GAAP for the S&P 500 companies

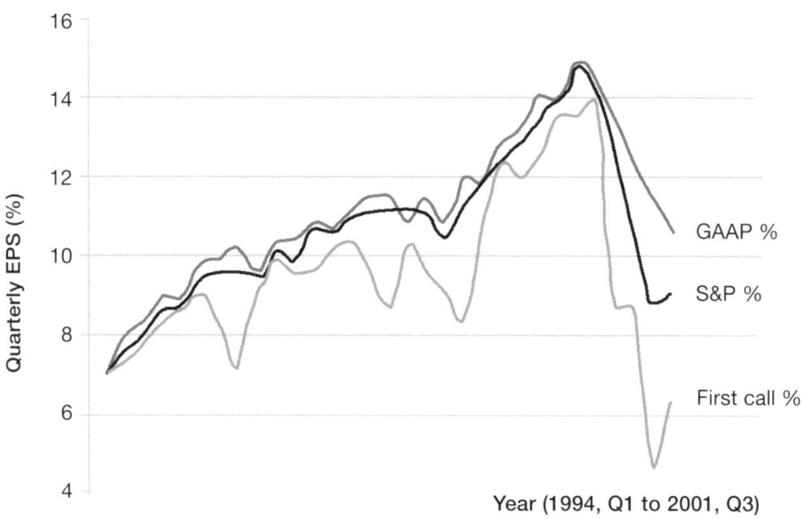

Source: Thomson Financial/First Call and Standard & Poor's.

Accounting's theory and practice are big contributors to the crisis. Both came of age in a more primitive business milieu. Like Newtonian physics, classical accounting worked well within the limits of a person's direct experience several centuries ago. Just as 20th-century physics was baffled by a slew of new observations and measurements in electromagnetism, chemistry, astronomy, and physics, so has classical accounting fallen prey to the confusion of, first, mass-manufactured value generation (with rifles and cars), and now post-industrial value (with software and knowledge bases). Physics responded with relativity and quantum mechanics; unfortunately, financial accounting has resisted any similar leap forward.

Perhaps the accounting area that became most abused during the 1990s' bubble economy was so-called pro-forma accounting. During this period, for every dollar of pro forma earnings that were reported, S&P 500 companies reported an average of 47 cents of earnings in their official GAAP reports a couple of months or so later, even excluding write-offs.[17] For the technology-dominated Nasdaq 100, a total pro forma profit of US$32 billion in 2001 became a GAAP loss of US$68 billion. Since then, critics have been quick to point out cases where pro-forma earnings have been used to inflate the bubble – a charge that gains credibility when you realize that if the S&P 500 was valued on GAAP rather than pro-forma earnings in 2001, its price-earnings ratio would go from an already expensive 25 times to 53 times.

In addition to pro-forma, a so-called thin float evangelism was used to boost prices; in retrospect, it has become known as the "dumb-investor effect."[18] High liquidity is traditionally seen as a sign of a healthy stock market, and indeed it is in an efficient market setting (a big if). But research revealed a "dumb investor" effect, where this extra liquidity was followed by a period of substandard performance. Baker and Stein[19] found that so-called dumb investors tend to be insensitive to market sell signals that the more informed investors pick up. Thus, they buy when others are selling, pushing prices up to unsustainable levels. But "dumb investors" can easily grow overly pessimistic, and then they simply withdraw, reducing liquidity. The effect is significant – when the turnover of shares is 12% higher than its long-term average, stock returns in the following year are 4% below their long-term average.

The fixation with pro-forma numbers can be traced at least back to 1982. Companies came under increasing pressure to produce the numbers that justified their rich valuations. An efficient yardstick was needed to provide comparisons between companies in each sector. The answer was pro-forma earnings – spiffed-up operating numbers that exclude many of the charges that reduce comparability between companies.

With valuations so high, any slips in earnings below expectations were punished with big selloffs in the stock market. As a result, much came to ride on stock prices, both as a currency in takeover deals and in the ability to keep key staff happy through stock options. Management realized they could not afford to miss expectations which often they were instrumental in creating. The charade worked in a bull market where fund managers played "beat the Street."

CRIMINAL NEGLIGENCE

Not all corporate crime is necessarily intentional. Some wrongdoing may arise from ignorance (easy when accounting rules grow overly complex), from over-ambition, or from opportunities taken when presented (right or wrong). Accurate accounting is instructional as well as preventive; it teaches right and wrong, and reminds individuals that wrong behavior will be immediately recognized and dealt with. Logical Positivists, a group of philosophers influential in the 19th century, held that if a statement could not be verified, it was literally meaningless. Worrying about them was simply flaunting one's confusion. Auditors should take this to heart – unintentional crimes are just as likely to occur (or perhaps more likely) because of fuzzy and incomplete accounting as from fraudulent accounting. Ethical violations don't require intent; negligence and stupidity are adequate.

Rules and abstract categories help in dealing with the real world – they allow us to get beneath the surface and ferret out hidden relationships that make a business tick. And because they are in a sense digital, they give representations stability and precision. If you make a map of analog copies from an analog cassette audiotape, the quality declines with each generation of the copying. But if you make a map of digital copies, the last can be as good as the first. Crisp symbolic representations allow for chains of reasoning in which the symbols are copied verbatim in successive thoughts, forming what logicians call a sorites:

- All SUVs are vehicles.
- All vehicle sales are profitable.

A sorites allows an analyst to draw conclusions with confidence, despite meager experience – absolutely essential if accounting is to be used by a broad audience with little common experience in a specific business model. For example, the analyst can conclude that SUV sales must be profitable even if the company has yet to sell an SUV, or if the first sale is projected for next year. The statement user can reach that conclusion even if he or she has never witnessed a sale of any automotive vehicle, but has only heard it described by a credible expert. If each step involves some projection on to a "fuzzy" monetary valuation (remember Plato's allegory of the cave), the slop accumulates. The last statement would be as noisy as an Nth-generation bootlegged tape, or as unrecognizable as the last whisper in a game of broken telephone. Yet this is exactly what we do in the accounting process called "consolidation." Analysts and investors need to be able to carry out long chains of reasoning built from links whose truth they could not have observed directly. (Otherwise, why have financial statements at all?) Philosophers have often pointed out that science is made possible by that ability. Without it, thinking grows decidedly unscientific. As Richard Dawkins observed:

> Airplanes built according to scientific principles work. They stay aloft, and they get you to a chosen destination. Airplanes built to tribal or mythological specifications, such as the dummy planes of the cargo cults in jungle clearings or the beeswaxed wings of Icarus, don't.[20]

Businesses, similarly, cannot escape the cold reality of science, despite the best arguments of hucksters and academic apologists.

Since the codification of double-entry accounting in Pacioli's *Summa* nearly 600 years ago, there have been many inventions which can improve on the information provided by financial accounting. Nineteenth- and twentieth-century developments in engineering and statistics have provided both a methodology for using existing qualitative and quantitative knowledge to predict the future, as well as a rigorous scientific basis for inference and understanding the businesses we are measuring.

The languages of engineering, statistics, and mathematics are powerful, since far-reaching implications can often be deduced from simple assumptions. Nevertheless, the function of mathematics must be seen as analogous to that of paint on canvas – it is only as good as the artist who uses the materials and the connoisseur who sees the result Without good underlying models, the professions of accounting, finance, and banking become little more than odds-makers for technology casino. The exploitation of new technology rests on sound science; so should investment in that technology.

The deterioration in "informativeness" of financial accounting has been a fairly recent phenomenon, dating only from the late 1970s and coinciding with the rise of information technology as the key enabler of business. This is hardly long enough to accommodate the glacial pace of rule-making at the FASB.

Up until about 20 years ago, both accounting and standard financial valuation approaches worked adequately well. What is different today? First, computers give us the capability of creating reports on-the-fly, and tailored to the specific needs of a decision-maker in a specific context. Or we can have computer algorithms – intelligent agents, so to speak – that can alert managers to information when it becomes available.

Second, computers and databases allow us to generate much more sophisticated statistics describing transactions than was possible with spreadsheet and ledger. For example, rather than the dollar amount of accounts receivable on a balance sheet, we might like the mean, standard deviation and count of individual accounts receivable transactions (from which we could recompute the amount if desired). Proper statistics, rather than mere account balances, provide analysts and investors with considerably more flexibility in generating analyses.

Critics of the traditional institutional approach to valuation – whether it is provided by accounting, or through methods based on accounting, such as discounted cash flow and residual income – cite four objections to the standard methodologies:

1. They are compromised by inadequate protocols.
2. They are too often applied in a context of flawed incentive systems.[22]
3. They provide no clear role for financial data, nor statistical protocols for predicting the future.
4. They do not provide a vehicle for routine learning that would encourage updating the forecasts and valuations each time new data arrived.

Traditional valuation approaches such as discounted cash flow and residual income rely on highly subjective choices for both data, and for the ways in which it structures these in the methodology (see Fig. 2.5). Such a "flexible" tool is flawed precisely because it can be used to contribute a quasi-scientific credibility to almost any valuation the analyst chooses in advance. But if the valuation is to be credible and scientific, then it must rigorously define a basis for judgments and decisions in each of the areas marked "subjective" in Fig. 2.5. It should also yield a single unequivocal valuation under a given set of assumptions – something at which traditional valuations fail.

Figure 2.5 Valuations are calculated from a mix of objective and subjective information

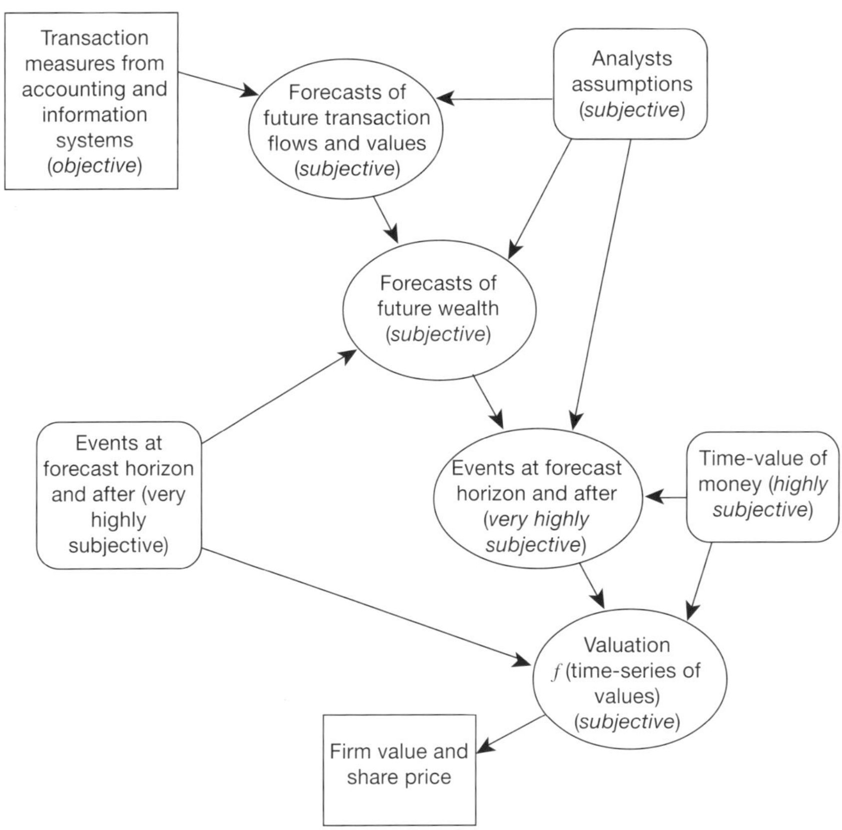

Flawed incentive systems in the audit and investment communities create the tendency to favor "rule-based" valuation systems, rather than valuation methodologies that would be most informative for their users and beneficiaries. Examples include annual audits of financials conducted by accounting firms which do most of their business in consulting (and thus may tend to use the rules to construct the valuations that the audited firms would like to see); or securities salesmen, who work up values to maximize trading gains and commissions.

In extreme cases of misreporting, companies are *ex post facto* required to restate their earnings to eliminate the most egregious misinformation. Restatements are now epidemic. In 2001, the amount of money that companies charged off annual earnings (a process which can severely obscure the real earnings from operations) reached a new high, as Figure 2.6 shows.

Figure 2.6 Annual charge-offs, including extraordinary and discontinued items

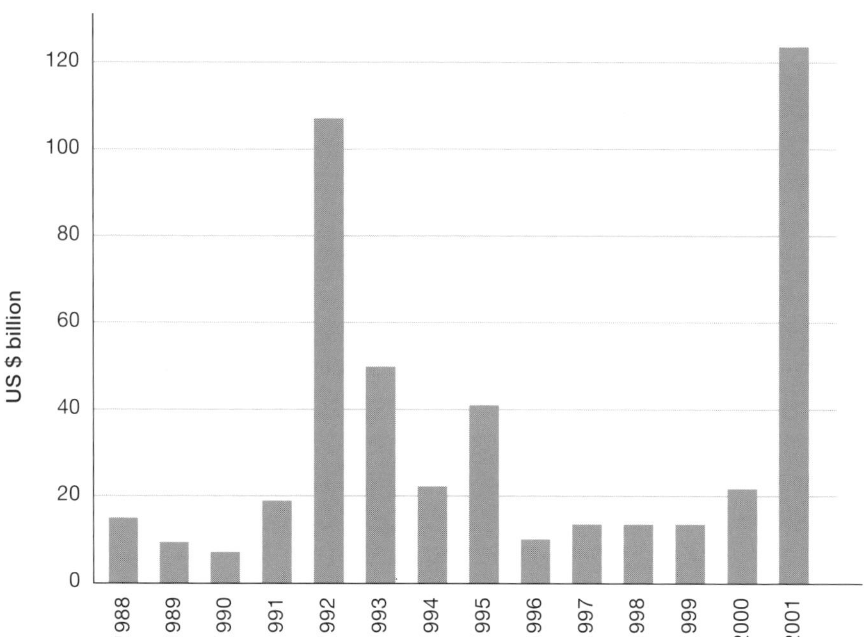

Even then, firms had difficulty deciding what their earnings really were – an astonishing 723 companies restated and lowered their audited earnings between 1997 and 2002. Charge-offs, restatements, and other accounting stratagems increasingly confound the forecasts of securities analysts, who may have biases toward overstatement to begin with. With enormous pressures to produce earnings growth, auditors are being turned into enablers.

The Enron debacle reinforces the stance of those who favor principles-based systems. They maintain that principles-based standards would have exposed Enron's problems long before the company went bankrupt, whereas GAAP's rule-based systems helped to cover up Enron's machinations.[23] While U.S. GAAP follows a cookbook approach – providing auditors with a comprehensive checklist of rules to follow – International Accounting Standards (IAS) is based on general principles, compelling auditors to enforce the spirit of the law and not just the letter. Harvey Pitt, former chairman of the SEC, has suggested that "rule-based" accounting provides "a road map for avoidance; if a rule says that above 10% an item should be reported, then those with something to hide will go for 9.9%."[24] He argues that under the principles-based accounting favored by international accounting standards, the Enron auditors would not just have looked at the numbers, but also would have quizzed management about why it had shunted assets and liabilities off its balance sheet, for example. And whatever management told them, the auditors would be required to make their own assumptions about what was actually going on. The Enron debacle clearly demonstrated the advantages of an objectives- and principles-based system as opposed to a rules-based system such as GAAP.

CAN ACCOUNTING IMPROVE ITSELF?

The short answer is, "Yes." Here is how. Virtually all valuation models – with the notable exception of historical cost-based financial accounting – are forward looking. In order to assess value, discounted cash flow, residual income (RI), real options, and other commonly used valuation approaches, predict the future value that the firm will generate from ongoing operations, strategy, and how it will play out given the market, technology, and competitive forces affecting the firm.

The accountants' balance sheet is, basically, a litany of costs sunk and obligations incurred in running the business in the past; income statements do little more than bridge from one balance sheet to the next. The accountants' perception of valuation is rooted in historical cost. The balance sheet's stored wealth perspective merely provides a tally of prior success or failures, with no indication whether these arose from a good business model, managerial vision, skill or luck, or from forces beyond managerial control.

An analyst's prediction, in contrast, is a summary of future activities, opportunities, and possibilities. Thus, accurate valuations (or at least the forecasts on which they are predicated) cannot be accomplished using backward-looking stored wealth models. They require process models which can effectively capture management plans for future operations. They need to be augmented by incorporating a time series of detailed transaction flows, or estimates from a time series of financial statements.

On top of these challenges lie the increasingly difficult challenge of ascertaining and managing risk. I would argue that accountants and auditors are in the best position to assess risk, because they have access to proprietary information that investors, banks, and securities firms do not. This is partially true, but we need to take into account the fact that auditors' and corporate accountants' domain is typically limited to a single firm. In practice, assessing the value of a firm must consider three realms of risk (Fig. 2.7):

1) risk from business operations;
2) risk from management's strategy and the way this plays out in the supplier and customer markets; and
3) risk in capital markets, which provide some of the financing for business operations (the rest being internally generated or from lending)

Figure 2.7 Three components of value; three contributions to risk

These three sources (or realms) of "value" in corporate valuation subject investors to a consequent risk of getting the valuation wrong. The firm's own business operations are under the control of management; vendor and customer markets, set the environment in which managers make their decisions; and capital markets, are subject to a variety of feedback loops and subjective biases in addition to fundamental business information.

The business operations realm is the realm over which management has the most control. It is also the only one of these three realms that accounting statements deal with. Where markets are stable, and the firm is competing effectively in the market with products that are accepted by consumers, then any changes in firm value can be assessed by looking at management activities in this realm. Unfortunately, ever since the commencement in the 1970s of "post-industrial society," as the sociologist Daniel Bell termed it, spot markets, outsourcing, rapid coordination of diverse producers, and so forth have increased the proportion of value contributed by the rightmost two boxes. The book-market ratio[25] of firm value has dropped from around 100% in 1980 to around 25% in 2002.

Significant changes took place in the second realm – that of vendors and customers – after 1980. Michael Porter of Harvard Business School recognized the impact of this second realm on value in the 1980s, and articulated their importance in value creation in two influential books – Competitive Advantage and Competitive Strategy. Management needs to set overall policy and strategy for their participation in end-product markets – that is, it must respond to the industrial structure and organization in their selected end-product markets. I specify end-product markets, because a company's products are the glue that binds its strategy. There are often multiple technologies for producing the same end-product – if you don't apply the most cost-effective technology, you will have competitors that do. Traditional accounting is completely silent on this realm, and the entity principle in accounting assures that any valuations by accountants will be restricted to business operations. But information technology, telecommunications, and global logistics have provided market alternatives that force much of the value creation outside the boundaries of the firm. Porter recognized this, but failed to describe specifically how the mathematics of value would work in this realm. Financial Dynamics offers an integrated option for assessing value from both business operations and competitive strategy in the industry.

The rightmost box in Figure 2.7 is the realm of the capital markets. These are markets for stocks, bonds, and other financial instruments that trade the firm's future potential earnings for cash advances. In this book, the operation of the capital markets is ignored, as that tends to vary in complex ways with the relative efficiency of particular markets (that is, ability to quickly and completely impact information into the market's prices) and in the liquidity, dispersion of beliefs (rational expectations), potential for fraud or misreporting of fundamental financial data, and other sources of error. Though these are important, they are considered outside the scope of this book. This book focuses on valuing business models, managerial effectiveness, and considering the interaction of those with vendor and customer markets. The current state-of-the-art in theory and practice in assessing value in capital markets is surveyed in MacKinlay and Lo's *A Non-Random Walk Down Wall Street* (the title a play on the title of Burton Malkeil's *A Random Walk Down Wall Street*)

Once upon a time, prior to banks, capital markets, and industry, accounting provided information on all three sources of risk. In those days, change was slow, inflation non-existent, and markets local, involving only a few tens or hundreds of people. Crop cycles dictated the creation of value, constrained by inputs. Predicting the future was easy; it looked like the past. Any unsold goods were simply stored until they sold or were self-consumed. Market revenues were costs plus a reasonable profit, which could be pushed to zero by competition (that is, cost = revenue). Count the inputs, and you had market value. Simple.

Today, accounting statements need to come to terms with risk and uncertainty. One of the fastest-growing job markets in the upper ranks of corporates is that of the chief risk officer. Debacles such as Enron Corp. and WorldCom Inc. are a sharp reminder to boards that assessing risk to a company's operations, market, credit standing, and reputation is essential. It is also essential in managing directors' insurance, which can be around US$50 million annually for US$200 million coverage. Here are some key factors that should be weighed in deciding how risky any business is:

- *Liquidity traps*: Even if a company can service debt, the way it manages its finances creates a set of separate risks. Moody's Investors Service said in a survey published in July 2002 that 771 investment-grade U.S. companies had disclosed only 22.5% of their triggers, some of which could force them to accept much worse terms on outstanding loans if their credit rating falls. Such triggers were critical in the failures at Enron and Tyco.

- *Materiality and dispersion*: Much of what is presented as hard information in financial statements is in fact a guess based on multiple, usually unstated, assumptions. Banks and credit-card companies, for example, set loan-loss reserves without revealing the assumptions they use to calculate them. They need to disclose such information and warn shareholders how their financial picture would change if their assumptions are off by, say, 1%, 5%, or 10%.
- *Internal control compliance*: Auditors provide "Management Letters," which assess the quality of their client's internal controls over business processes, and internal audits to assure that these processes are actually operating as management thinks they are (and as the financial accounts report that they are operating). Shareholders need that information, but they don't get it because companies aren't required to reveal it.

Former Andersen CEO Joseph F. Berardino asserted three ideas worth considering in light of Andersen's collapse:[26]

1. Rather than a pass–fail system of approving financials, auditors should hand out report-card grades that reflect the quality of a client's accounting. That would give auditors more leverage over clients in accounting disagreements and also enable the market to give a premium to companies that get the highest marks and discount those that get the lowest grades.
2. Change the focus of directors so that they are involved in risk management, not the verification of financials. The most important questions that a director can ask are: "What are the most significant risks to this company? How likely are they to occur? What is management doing about them?"
3. Change the financial-reporting system so that it is based less on "a game of rules, loopholes, and legalisms" and more on principle-driven accounting – the approach of the International Accounting Standards Board.

In a nutshell, Berardino is telling us that we need accounting systems that embrace uncertainty and create a clear role for risk. Only then can we use them to manage the complex and increasingly uncertain business of the 21st century.

ACCOUNTING AND ETHICS

The most important contribution of good accounting to business is ethical. You cannot manage what you cannot measure. By making the inner workings of the firm visible to all parties – not only investors, but also employees, customers, and suppliers – pressure is appropriately placed to force companies to act ethically in the interest of all stakeholders (not just shareholders and top executives). Accounting promotes accountability, fostering an open corporate culture where ethical lapses, unfair practices, and antisocial behavior are easily reported and fixed.

If corporate accounting is truly the first step in all of our measures of economic performance (and I think there would be widespread agreement on this point), then poor accounting naturally invites – at best – poor management; at worst, it invites corruption as ethical employees and managers bail out of the system. When investors have their confidence bashed by corporate malpractice and number fiddling, almost any half-suggestion of impropriety can make them twitch, running stocks down irrationally.

Better accounting does not mean more numbers – the interpretation that corporations and accountants have often applied. Quality of information counts, not quantity. And quality implies that the appropriate measurements for the appropriate decisions are given to the appropriate users. We will see where this concept takes us when we discuss the Real-Time Corporation in Chapter 12.

The U.S. experienced such a loss of faith in the 1930s when, in the aftermath of a stock-market bubble and the 1929 crash, Wall Street became public enemy number one. Business went into a severe decline as investors and companies lost faith in the trading and issuance of securities. The Pecora hearings in Congress provided an outlet for the public's Depression-era fury and resulted in several laws that imposed unprecedented regulation on the business of finance. The creation of a "cop on the corner" of Wall Street – in the form of professional accounting and auditing, and the SEC – was a stroke of genius, greatly increasing public faith in America's capital markets, maximizing liquidity, minimizing manipulation, and assuring the middle class's enduring participation in corporate ownership.

That faith came increasingly under attack in the 1980s and 1990s as the portion of wealth audited and tracked by the accounting profession dropped from around 100% in 1980 to around 20% in 2002. With so much corporate activity unmonitored and unreported, it was only natural for investors to become easily spooked; and, unchecked by public scrutiny, for the most renegade and capricious of managers to grow powerful.

Monitoring and enforcement, detection and correction, policing and punishment, auditing and wealth redistribution are contrasting approaches to the same end: to keep bad things from happening to your system. The first is unintrusive, and easily accepted by the populace or constituency; the second is draconian. The second is hard to get right if you must "prove beyond any doubt" that a party is guilty. Audit and transparency are fundamentally preventive; their mere existence causes people to self-enforce their behavior. That is why auditing and accounting are important.

The consequences of the late-1990s' bubble have been dramatic. Household saving, in particular, has declined as Americans reassess what they can expect from their retirement portfolios, and has led to sluggish consumer spending. Capital investment has suffered and firms have become cautious about borrowing, stifling innovation.

During the dot-com boom, Merrill Lynch's Internet guru, Henry Blodget, abused Internet firms with various expletives, while simultaneously issuing research reports that urged investors to buy their stock. Such duplicity was singularly possible because financial accounting was at its worst in describing the value-generating potential of knowledge-intensive companies such as the dot-coms that made Blodget rich. This is largely due to failures of accounting and corporate governance that, although not really Wall Street's fault, have been used by Wall Street to hide the truth – such as at Enron, with its various off-balance-sheet financial structures.

Blodget was able to hide behind the pretext of *caveat emptor* (let the buyer beware). Yet such an excuse is credible only if the investor can go elsewhere to find information – for example, in audited financial statements. On June 3, 2002, a Supreme Court ruling stated that the securities markets' regulations introduced in the 1930s "sought to substitute a philosophy of full disclosure for the philosophy of caveat emptor, and thus to achieve a high standard of business ethics in the securities industry."

The point is clear. Ethics – both Wall Street's and corporate ethics in general – suffered dearly as a direct result of the breakdown of financial accounting for knowledge economy firms. The first step in renewing corporate ethics is to update financial accounting, and its audit processes, to meet the dictates of the knowledge economy. Any suggestion of conservatism in meeting the challenge of the knowledge economy can only be interpreted as an invitation to further corruption and volatility.

---

1 Plato, *The Republic* (London: Penguin, 1968) (abridged by the author).

2 "The Army's Secret Weapon: Repair Shops in the Field," *Business Week*, July 22, 2002, p. 63.

3 R. Dawkins, *The Selfish Gene* (Oxford: Oxford University Press, 1976), p. 192.

4 D. Dennett, *Darwin's Dangerous Idea* (London: Penguin, 1995), p. 365.

5 D. Dennett, *Consciousness Explained* (Boston: Little, Brown, 1991), p. 202.

6 Dawkins, op. cit.

7 Financial Accounting Standards Board, *Statement of Financial Accounting Concept No. 1: Objectives of Financial Reporting by Business Enterprises* (Stanford: November 1978).

8 George Gilder, "Introduction," in Sexual Suicide (New York: Quadrangle, 1973).

9 Friedrich August von Hayek, The Counter-Revolution of Science (New York: Liberty Fund, 1952), Part 1, Chapter 3. Von Hayek's ideas form the basis for the so-called Chicago school of economics, which has defined most of the focus and direction of academic research in accounting and finance since the 1960s.

10 "The Same – Only More So?," The Economist Technology Quarterly, December 8, 2001, p. 10.

11 Raymond C. Kurzweil, The Age of Spiritual Machines: When Computers Exceed Human Intelligence (Penguin: New York 2000).

12 Joseph Berardino, 2001, quoted in Learing From Mistakes, *Business Week*, p. 42, Aug. 12, 2002

13 Though it should be noted that, for a period of time in the 1960s and 1970s, Arthur Andersen offices conducted external audits of the income statement, but were ultimately forced to switch to auditing the balance sheet as rules changes made auditing more competitive and cost-conscious in the 1980s.

[14] See E. Ziglio, "The Delphi Method and its Contribution to Decision-Making," in M. Adler and E. Ziglio (eds), Gazing into the Oracle: The Delphi Method and Its Application to Social Policy and Public Health (London: Jessica Kingsley Publishers, 1996), p. 33; J. G. Wissema, "Trends in Technology Forecasting," R&D Management, 12(1), 1982, p. 27–36.

[15] B. Nussbaum, "Can You Trust Anybody Anymore?," Business Week, January 28, 2002.

[16] Quoted in "Special Report: The Trouble with Accounting," The Economist, February 9, 2002, p. 61–63.

[17] John Ogden, "Pro-forma Fantasy," South China Morning Post, February 27, 2002, B2.

[18] "Special Report: The Crisis in Corporate Governance," Business Week, May 6, 2002.

[19] Malcolm Baker and Jeremy C. Stein, "Market Liquidity as a Sentiment Indicator," NBER Working Paper #8816 (AP, CF), March 2002.

[20] Richard Dawkins, *River Out of Eden*, (New York: Basic Books, 1995), Chapter 2.

[21] As measured by Book÷Market value; not a perfect measure, but if investors are not systematically overpaying for equities, then it must be reasonably precise.

[22] "A Penny in Whose Pocket?," The Economist, May 26, 2001, p. 75.

[23] David Fairlamb, Carol Matlack, and Moon Ihlwan, "Europe's Bean Counters Are Sneering," Business Week, January 28, 2002.

[24] Quoted in "Special Report: The Trouble with Accounting," The Economist, February 9, 2002, p. 61–63.

[25] This is often used as a surrogate for Tobin's Q, since it can be computed from observable measurements.

[26] J. A. Byrne, "Fall from Grace," *Business Week*, August 12, 2002, p. 16.

CHAPTER 3

Ideas in Motion

What impresses men is not mind, but the result of mind.

– *Walter Bagehot*[1]

WHAT ARE EARNINGS? "STORED WEALTH" AND "VALUE FLOW" IN ACCOUNTING

How should we define "earnings" – a concept so essential to modern business, yet so elusive? Fifty years ago, the financial accounting community still believed in a "true income" property of individuals and firms. From their perspective, "true income" held for the firm the same essential validity as a physical constant such as Planck's, Avogadro's, or Boltzmann's constant. All that was needed was proper refinement of accounting systems to discover it.

Subsequently, the profession and its regulators discarded any illusion of a "true income" and accepted the reality: that income is a more or less arbitrary and inconstant measure which needs to be constantly redefined by the profession as the business environment evolves. This is unfortunate, because, in a sense, accounting lost its anchor; without an overriding commitment to a particular objective, accounting has tried to become all things to all people. If "true income" was an overly ambitious goal, the implicit objective of satisfying all potential users of financial statements with one simple approach must be impossible.

We have seen the results of this overreaching. During the bubble years of the 1990s, it depended on what a company's management decided to throw in and leave out of its pro forma earnings results. Interestingly enough, the discrepancy between "reported" and "operating" earnings only became significant during the decade of the 1990s, when almost every investment went up and most investors cared little about any form of earnings. What was dearly needed during the dot-com bubble was a set of rules for calculating operating earnings and a requirement to make clear how these calculated numbers relate to the GAAP net income. In the end, investors need two numbers – a standardized operating number and an audited net-income number – and a clear explanation of how to get from one to the other.

With some controversy and considerable fanfare, Standard & Poor's, the credit rating agency, in 2002[2] brought out new standards for calculating operating earnings figures (though companies are not required to act on S&P's proposals). S&P's new standard treats employee stock options as a quarterly expense against earnings. That particularly hampers technology companies, which often issue stock options in lieu of salary. The stock options change alone cut 2002 earnings for companies in the S&P 500 index by an average of 10%. Other items included in operating earnings by S&P are restructuring charges, write-downs of assets which are depreciable, and the cost of buying outside research and development services; items excluded are pension-plan investment gains, unrealized gains and losses from hedging activities, and merger-and-acquisition related expenses.

All future predictions are uncertain. I only remind you because an essential part, and the most difficult task in assessing firm value, is that of predicting future value flows for the firm. These predictions will contain error, and you would like to have some idea of what the magnitude of error is, and whether there are behaviors and relationships in this error about which you can learn. If you have the choice, you would also like to minimize that error, or make it obey some other sort of regularity constraint. The goal is consistent estimates based on a clearly articulated set of (1) future strategies, and (2) projected environmental conditions – including market, technological, ownership, and decision option situations. This will minimize, but not eliminate, uncertainty. It also defines a clear path to redefine estimates as events unfold in real time, and true market and environmental conditions reveal themselves over time. Because in the knowledge economy, earnings are the consequence of ideas in motion.

HIGH CRIMES AND DOUBLE-ENTRY

On June 27, 2002, markets around the world were sent reeling when it was discovered that WorldCom had overstated the prior 15 months of earnings by US$3.9 billion (to which was later added another US$3.2 billion for a total of US$7.1 billion in accounting misstatements). For a company that reported US$1.4 billion net income in 2001, it would be difficult to dismiss this as "immaterial." Roman Weil, a professor of accounting at the University of Chicago, noted that WorldCom's fraudulent accounting "is so basic that I teach it in the second week of my class." Yet the ploy, which misclassified supposedly difficult-to-manipulate cash flows, fooled both Arthur Andersen and KPMG, two of the (at the time) Big 5 accounting firms. "How do you fake cash flow? You simply move the negative things – the cash outflows – out of the operating section and you move it into the investing or financing section. It's exactly the thing that WorldCom did," noted Howard Schilit, head of the Center for Financial Research and Analysis. What was significant was that few companies used the stratagems that undermined Enron; but all corporations use cash flow and earnings before interest, taxes depreciation, and amortization (EBITDA) as a measure of value. And cash flow has been championed by the analysts' community that claims that it is not subject to the ambiguities of "income." Arthur Bowman, editor of Bowman's Accounting Report, noted that "that huge amount of money would involve transactions that would have to be justified," concluding that the paucity of Andersen's and KPMG's statistical sampling failed dismally in identifying a fraud that was not just material – but which the SEC decided was criminally negligent. (I will speak at length on the failings of current audit approaches in Chapter 11.) Accounting fraud allowed WorldCom to raise about US$12 billion in capital in 2001 from the sale of debt, largely because investors believed it was stronger than competing telecommunications start-ups with real assets and healthy levels of cash.[3]

Did generally accepted accounting principles (GAAP) contribute to the fraud? Yes; indeed, GAAP is a prime enabler of fraud. Without double-entry bookkeeping, frauds such as WorldCom's could never be perpetrated. Here is why. The technology of double-entry bookkeeping was developed in the 12th century, and codified by Luca Pacioli in the 15th century, at a time when people who could read and write were scarce, and those who additionally knew arithmetic were even scarcer. There was no computing equipment – indeed, Pope Sylvester II had been accused of being in league with the devil simply because he championed the abacus. In this rudimentary environment, a system that counted everything twice provided about the only control over accuracy. Double-counting survives to this day as a cornerstone of accounting, even though the universal adoption of computers for accounting relegates error control to computer input. So, instead of error control, double-entry offers numerous opportunities to misstate financial transactions. Transactions start off as "journal entries" in the accounting system, which usually follow the general format set out here:

Debit: real (flow) account	or	Debit: nominal (stock) account
Credit: nominal (stock) account		Credit: real (flow) account

All of the journal entries affecting a particular account are accumulated, and the total appears as the account balance. Real accounts represent rates of flows of value – for example, sales for the year 2002, whereas nominal accounts represent stocks of something stored at its historical cost – for example, inventory at December 31, 2002. The terms "debit" and "credit" were medieval terms that meant "debt" and "to entrust," respectively; this reflected their maritime trade and agricultural usage. Today they are dispensable vestiges of a simpler era. Managers, investors, and analysts are interested in reconstructing the flows of transactions throughout the year at the firm level. Instead, the accounting tradition discards the original transaction information, and summarizes them into the arbitrary classification scheme (current assets, property, debt, and so forth) that applied well to 15th- and 16th-century agriculture and trade, but which has grown largely irrelevant in the knowledge economy.

IN SEARCH OF ACCOUNTING'S OWN "MINKOWSKI SPACE"

At the turn of the 20th century, the mathematician Hermann Minkowski proposed what he called his "world postulate" – that the real world exists as discrete "events" in a continuous and unified space-time (Minkowski Space). This perspective was to have a profound influence on 20th-century physics, as it provided the modeling context for Einstein's general theory of relativity. According to Minkowski, constructs such as time, measurement, and location in space, and various other metrics which had been used to express Newtonian physics, were merely projections of reality into computationally convenient frameworks that excluded one or more aspects of reality.[4] In physics, models of reality had to be revised – resulting in relativity and quantum mechanics – when empirical results showed 19th-century Newtonian physics was flawed or incomplete.

Financial reality also plays out as a series of economic events in a continuum. Financial reports provide a particular context for summarizing transactions – that is, discrete economic events that take place in a "taxonomy-value-time" space. A transaction takes place at a particular point in time, and involves a given value flow of significance to some firm, within a particular class of activities. Financial reporting systems make implicit assumptions about what computer scientists call "granularity." Traditional accounting classifies events in time periods of one year; it uses taxonomical classifications – for example, assets, liabilities, debits, credits – that arose from medieval trading; and it uses discrete values measured in multiples of a not necessarily invariant monetary unit – for example, one dollar. There is a tendency to treat each of these conventions as real – for example, we often speak as though "income" is a real attribute of a firm, whereas it is merely a stylized measurement resting on the traditional accounting taxonomy. As the business world has evolved, empirical evidence increasingly points to the flaws in the traditional "granularity" of financial reporting. Most damning is the fact that financial accounting seems now to describe less than one-fifth of the value of the firm. In accounting, our models of reality have yet to be revised, even though empirical results have shown 14th-century Paciolian accounting to be flawed and incomplete. Financial accounting has yet to find its Minkowski.

The entity concept in GAAP accounting dictates that economic events only include those happening as "arm's-length transactions" with outside parties – there needs to be a clear transfer of ownership. This is a fairly crude measure of value flow, which must have worked well for the Venetian merchant marine, which only traded commodities on consignment. But it is grossly inadequate for describing business in the knowledge economy.

Consider the example of Microsoft's Office Suite, which is one of Microsoft's "cash cows". If we consider value only to flow from "arm's-length transactions" with outside parties, then we will incur a long period of costly programming and debugging (where we pay our programmers and vendors at historical cost), followed by a period of nearly pure revenue at the market price (since we can duplicate and distribute the code almost costlessly). Neither carries their "stock" of Office software as an inventory item, nor anywhere else as an asset. From the traditional accounting perspective, there is no difference in value between Microsoft's Office Suite and Corel's WordPerfect Office. Even though they are functionally the same, most financial analysts would place a much higher value on Microsoft's Office Suite. Why? The standard explanation is that expectations of future revenue flows are higher for Microsoft than for Corel; but, of course, this has little to do with historical "arm's length transactions." No wonder accounting leads analysts to shameless value flogging and loss of objectivity; financial accounting fails to provide useful information on value processes at Microsoft and Corel.

Histories of economic events provide some information that is useful in predicting future firm performance, but they are not the sole determinant of future value. Furthermore, the summarizing of economic events into a standard set of account classifications discards information about the processes that drive value creation. Much of the published financial analysis is directed toward extracting a true history of economic events from the distorted summaries provided by this standard taxonomy of accounts. WorldCom's misstatement of cash was possible because there were several alternative accounts into which what was basically a cash transaction could be disposed of. Wouldn't it be more useful just to summarize the total value and timing of cash flows in a single "cash flow" rate? More fundamentally, economic events (that is, accounting transactions) can be defined by two parameters and a descriptor – for example, a cash disbursement transaction might be defined by the triplet {cash, -1,000, June 1, 2001} – and a set of these could be presented either as a flow rate – for example, {cash, -10,000,000 per year} – or a stock value – for example, {cash, 1,500 at December 31, 2002}, or both. This ostensibly is what the combination of income statement and balance sheet should do. But any information is confounded and muddied by double, quadruple, and multiple counting. This is why statistical studies using cross-sectional financial data often have very low goodness of fit (for example, R^2 and F-statistic). A cleaner approach would simply report flow rate statistics of each particular class of transaction relevant to the business model, and report it directly, without worrying whether assets = liabilities + owner's equity, or debits = credits. These artifacts of medieval Europe have outlived their usefulness and should be discarded.

Managerial strategy, policy, competitive forces, supplier and customer relationships, patents, copyrights, and other claims to property all contribute to future performance – and thus to firm value. Any proper accounting system needs to consider these "policy" and "environmental" variables, as well as the purportedly "objective" histories of arm's-length transactions which are at the core of traditional GAAP.

VALUE, INFORMATION TECHNOLOGY, AND THE KNOWLEDGE ECONOMY

The Industrial Revolution was made possible by machines that would allow one person to do the physical labor of many. The knowledge economy was similarly made possible by machines that would allow one person to do the mental labor of many. In neither revolution was there a smooth transition; rather, there were false starts, dead ends, and opportunities missed or delayed. Retooling took years, and gaining potential efficiencies decades. This made them no less real. We have had two centuries to reflect on the Industrial Revolution; we have been reflecting on the knowledge economy for less than two decades, and are immersed in developments brought about by information technology. Yet, over one-third of the acceleration in labor productivity in the late 1990s when compared to the preceding 20 years was due to information processing improvements in output of machinery, electrical equipment, communications and a software.[5] Indeed, productivity growth can be tracked to show which technologies contributed to the revolution and when. Clear evidence of the revolution in productivity can be gleaned from figures derived from the US national income and product accounts

(NIPA) for "information processing and software" which can be expected to grow in tandem with the role of knowledge in producing wealth (Fig. 3.1 and Fig. 3.3).

Figure 3.1 Information technology contributions to real GDP growth

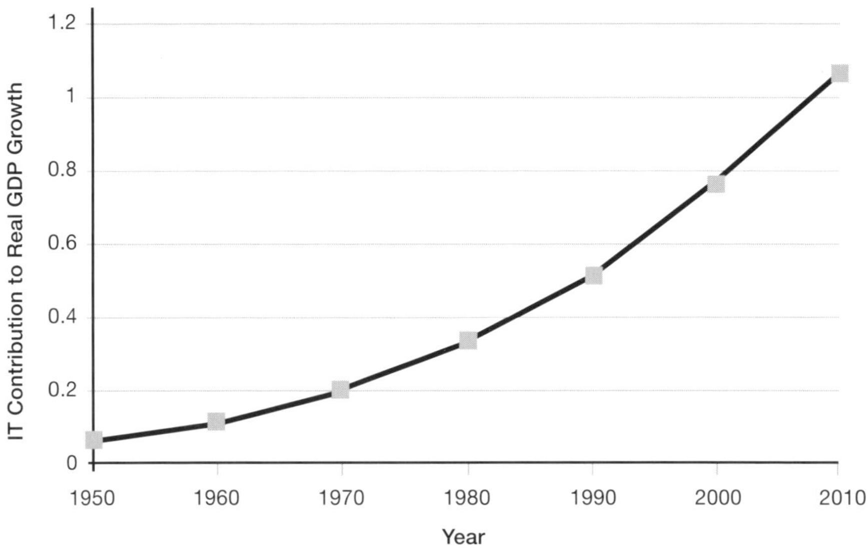

Source: U.S. national income and product accounts for information processing equipment and software, 1951–2001; 2002–10 projected.[6]

There are various reasons to believe that figure 3.1 may understate the importance of information technology on economic growth. Many goods and services produced using high-tech capital may not be adequately measured, which would help to explain the surprisingly low productivity growth in many of the high-tech-intensive service industries.[7] Measurement problems are more widespread in national accounts than is often admitted. For example, 23% of GDP is measured using either physical inputs as extrapolators (mainly labor hours) or as input-cost indexes, which produce zero or low growth in labor productivity and often negative growth in multifactor productivity because of the rapid rate of growth in investment and capital stocks. Furthermore, information on wages and salaries is collected only for production and non-supervisory workers, thereby omitting nearly 50% of employee compensation.[8]

During the two decades leading up to the widespread use of the Web, starting in the summer of 1995, U.S. labor productivity rose on average 1.4% annually. From 1995 to 2001, it rose 2.4% annually, and is projected to accelerate to 3–4% annual increases in productivity in the coming years (although this number, like any forecast, is subject to considerable debate). Figures 3.3 and 3.4 show the trend.

The contributions of information technology to the U.S. economy can be divided into several periods (Fig. 3.1):

- Embryonic (1950–64): This is a period when computers were little more than calculators, useful for the specialized computations of insurance companies and other "number crunching" applications.
- Mainframe era (1964–80): The defining moment of the mainframe era was the introduction of the IBM 360 on April 7, 1964. Nearly two decades earlier, Thomas Watson, Sr., CEO of IBM, claimed there was a market for perhaps only five computers. His son Thomas Watson, Jr. became IBM's CEO in the 1950s, immediately embracing the new computing technology even though it cannibalized existing products. The "360" alluded to IBM's trademark "full circle of service" providing software, on-site support, and other necessities for the first time in the marketing of computers. But software was not seen as a separate market; rather, it was seen as a "loss leader" necessary to sell IBM's hardware.
- Mainframe software (1970–80): Database manager DL/1 (acquired from North American) and network managers such as CICS (acquired from Union Carbide) became "killer applications" on the IBM mainframes, and were used to aggressively market IBM's hardware. These were primarily accounting machines, and their implementation resulted in the reduction of bullpens of thousands of accountants to perhaps 10% to 15% of their original number. Those that remained were less concerned with accounting than with the care and feeding of the machines – that is, data input and output.
- Microcomputer era (1980–95): The defining moment of the microcomputer era was the introduction of the Apple II on April 1, 1980. The development of killer applications such as spreadsheet processors, as well as simplicity and low cost, made them an immediate force in the workplace, especially once IBM entered

the market with the PC. Data rapidly dispersed away from the mainframe databases, and on to the desks of employees who were thus "empowered." One of the less savory consequences of this empowerment came to light during the "junk bond" scandals that resulted in the market crashes of 1987. Securities salesmen found that, rather than having to depend on research departments for their analyses, they could prepare them on the fly with PC spreadsheet processors; and, conveniently, could force whatever results were required to make the sale. The evolution of local area networks throughout this period (in the 1980s to share expensive disk space, but later to share the data on the disks as disk storage prices plummeted) increased the utility of PCs until, in the early 1990s, PC spending outpaced mainframe spending.

- Web era (1995–present): The Internet was commercialized in 1991 (when the dot-com was born), and the World Wide Web came into being in May 1993. But it wasn't until Microsoft embraced the Internet in June 1995, redeploying 10% of the workforce to Internet projects in the process, that industry took note. Later in the summer, attempts by Microsoft to enter Internet banking through the acquisition of Intuit in July 1995, and later still, the successful IPO of loss-making Netscape in August 1995, set the stage for the dot-com boom and for an explosion in personal as well as business Internet use. Corporate data soon became widely available in an interactive form. Forward-looking firms such as Dell Computer built successful businesses around this immediate global access to individual consumers.
- Embedded systems era (the near future): The embedded systems era has begun, though mainly unnoticed. Robotic toys, electronically managed automobile engines, smart telephones, and numerous other appliances have grown to become the dominant consumers of chips and computing power. More importantly, the knowledge economy is getting legs. Formerly paraplegic computers, whose feeble input and output were solely limited to keyboard and screen, are now ambulatory; they have electronic noses and taste buds, can hear and understand sounds, and can speak. The impact of the microcomputer will pale in comparison to sentient, robotic computers.

Of course, there is room for error, since equipment and packaged software are to some extent the "tail wagging the dog" of automated information. First was the transition of archived information (for example, government and corporate records, and so forth) off of paper and on to computer databases, and the transition, with cheap, high-quality laser and dot-matrix printers, of paper toward being a medium solely used for displaying information. The next transition, perhaps in a decade, will eliminate paper as a display medium, replacing it with cheap, flexible, high-resolution display screens and "digital paper." Such a medium will invite reporting that will update on-the-fly to meet the immediate needs of the reader.

Do we really want to encumber this with accounting methods that stabilized before the invention of movable type?

In addition to the standard challenges associated with measuring real output in these statistics, measuring real output in information technology presents additional challenges because new products are constantly developed and introduced and because existing product characteristics tend to change rapidly – the technology accelerates.[9] These challenges will be met in Financial Dynamics through the development of information technology-specific hedoninc price indexes and techniques that take into account the value of ideas.

GAAP's model of the firm's business is an exercise in studied ambiguity – businesses are modeled in terms of a murky flux of formal and informal processes that generate stored wealth (that is, assets) while using up resources (that is, costs). This was perfectly adequate in the agricultural economy for which it was developed. Financial Dynamics focuses, in contrast, on the processes that generate wealth. This makes much more sense in the knowledge economy, which has knowledge assets that are difficult or impossible to store, which quickly grow obsolete, and which (because they are non-rival) may be impossible to unequivocally own or protect from theft.[10]

Over the past two decades, knowledge-intensive and asset-lean industries have moved from the periphery to center-stage, now generating 80% of the wealth and employing 85% of labor in the U.S. and other developed economies. Firms that invest heavily in physical assets are the firms earning the lowest returns. Figure 3.2 shows that shareholder return and asset intensity (as measured by the ratio of fixed assets to sales) are inversely correlated. Microsoft and its products are a primary example of this transition. In the late 1970s, Bill Gates was ridiculed for building a company to produce software. Everyone then knew that software only existed to sell more hardware (the IBM model of the day). Two decades later, Gates's insight is received wisdom to the point of antitrust lawsuits. The point is clear: value can no longer be locked in a vault and hidden away under the collective mattress. Value has evolved to become process- and knowledge-based. And the $10 trillion U.S. economy now drives the pace of the rest of the world's US$30 trillion economy.

Figure 3.2 Shareholder return and on-balance-sheet fixed asset intensity, 2000

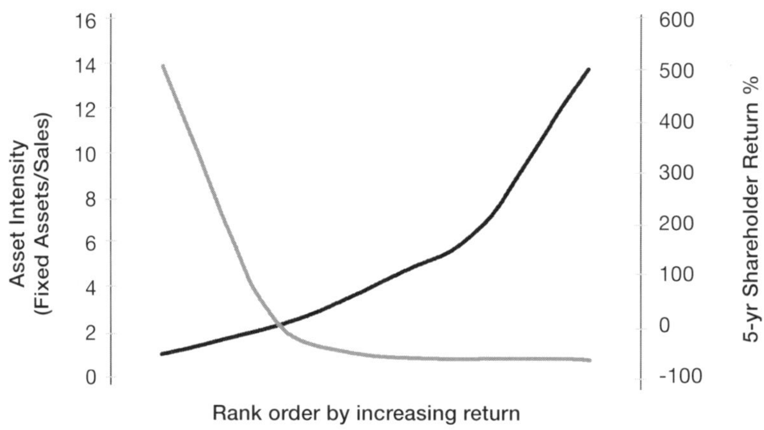

This phenomenon is so recent that the accounting and finance communities are still coming to terms with its ramifications. We are emerging from a long tradition of wealth measured in terms of storable assets. The history of money is a history of storable commodities – salt, leather, silver, and gold. The great contribution of Adam Smith's The Wealth of Nations was to soundly refute Thomas Mun's arguments that wealth could only be measured in terms of storable commodities such as gold and silver. Smith's inspiration came from the colonial Americans and their energetic redefinition of wealth in terms of their own labor and innovation. Two centuries later, the production of tangible assets employs less than 15% of the U.S. workforce; and the value of these assets is less storable than ever. Think about a personal computer – its value is mostly used up in two or three years, not because of wear, but because of technological obsolescence.

Since Croesus of Lydia standardized gold coins, making wealth both universal and tangible, money has moved from being a vehicle for tributary control to become a vehicle for personal expression. With it, mathematics and accounting moved from the realm of magic, into the everyday world of the merchant. But tangible wealth has always had problems as well – gold- or silver-based currency is limited by the amount of gold or silver in the world. The market fluctuates with each new discovery, and each new technology development (a development which destroyed the economic supremacy of the Spaniards after the 15th century). Paper money solves these problems, since supply can be modulated simply by adjusting the speed of the printing presses, as well as divorcing users from the myth of a physical basis for wealth. Adam Smith argued even further, that wealth could be valued in terms of the productive vitality of an individual, firm, or nation.

The word value has two different meanings, and sometimes expresses the utility of some particular object, and sometimes the power of purchasing other goods which the possession of that object conveys. The one may be called "value in use"; the other, "'value in exchange." The things which have the greatest value in use have frequently little or no value in exchange; and, on the contrary, those which have the greatest value in exchange have frequently little or no value in use. Nothing is more useful than water: but it will purchase scarcely anything; scarcely anything can be had in exchange for it. A diamond, on the contrary, has scarcely any value in use; but a very great quantity of other goods may frequently be had in exchange for it.[11]

Adam Smith's economics favored a concept of wealth created by productive labor – a labor theory of value.[12] In his theory, value is the amount of labor that will be traded for a good or service. In practice, such a definition requires accurate predictions of future activities. The labor basis for valuation has grown tricky in a world of rapid automation with accelerating technological invention. The conundrum is that if we can produce precisely the same product tomorrow that we did today, but with half of the labor, then how should we price our inventory. From either the exchange or the production perspective, we have a value that changes under different assumptions about future managerial policy and strategy. The impact of automation, as well as other strategic decisions, must therefore be explicitly incorporated into our valuation models. I will show how this is possible when we cover prediction in the Financial Dynamics methodology.

Accounting which tracks assets only as storable commodities at historical cost misses most of the value creation going on in the 21st century. Perhaps this is why accounting now measures only a small percentage of corporate value. For this reason, too, I am uncomfortable with most of the newer "stored wealth" approaches currently used to value so-called knowledge capital. "Stored wealth" approaches attempt to classify particular historical costs for patents, software, and so forth. Such approaches are reflected in the software and goodwill accounting regulated by FASB 86 and FASB 141[13] and is reflected in various "brand equity" formularies now popular in marketing.[14] Columnist and management writer Tom Stewart summarized in his book *Intellectual Capital: The New Wealth of Organizations*[15] various definitions of intellectual capital – that is, knowledge, information, intellectual property, experience – that can create future wealth. But a "stored wealth" definition of intellectual capital is, at best, a slippery concept, covering everything from a patent granted on a new type of gadget to the personal relationship between a top salesman and his best customer.

The real problem lies in measurement and, in consequence, in the strategic use of any derived valuation. The knowledge economy does not see production in terms of lumps of value (like lumps of gold) that sit around forever until sold. Rather it adopts a "value flow" perspective on valuation makes the intrinsic assumption that information, knowledge, and other intangibles only generate value through the use of ideas in the revenue-generating activities of the firm. Traditional notions of discounting future revenue streams to compute value, or matching expenses to ongoing activities, are perfectly applicable in this scenario.

The process perspective presumes a clear understanding of how the particular knowledge assets in which the firm invests can in fact be used in the future to generate value. Since there is no certainty in future corporate action, contingent scenarios (such as those addressed in real options and decision tree valuations) need to be considered in calculating value.

In many ways, strategy and operations have been suggesting these additions for some time. Reengineering described how technology redesigned jobs and relationships in the firm. Core competencies were various activity sets, areas of expertise, and so forth, around which particular value chains were organized. And all of this is increasingly influenced by the external political, market, and social climates, which are not even considered at all in traditional accounting.

Figure 3.3 Proportion spent on particular IT components of information processing in the United States

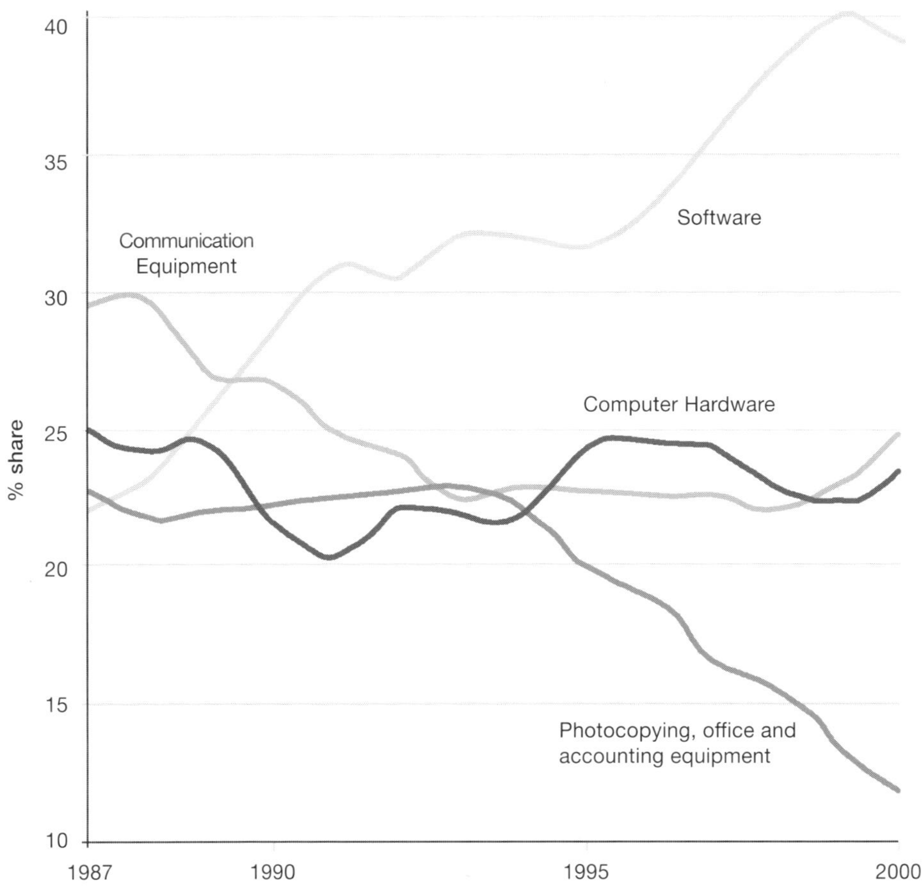

Figure 3.4 U.S. labor productivity percent increase on year before (1990-2005)

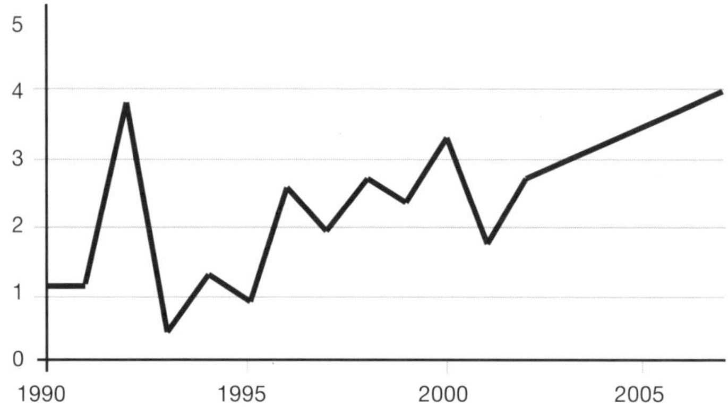

In the next four chapters, we will look at the ways in which these novel and problematic attributes of knowledge intensive technology businesses are addressed with Financial Dynamics valuation methods.

---

1 W. Bagehot, *The English Constitution* (1867 reprinted March 2001; Cambridge: Cambridge), Chapter 8.

2 These were laid out in a memo circulated to the press in November 2001.

3 "Curiouser and Curiouser," *South China Morning Post*, June 28, 2002, Business 2, p. 1.

4 See Minkowski's address in A. Einstein, *The Postulate of Relativity* (New York: Dover, 1952).

5 S. D. Oliner and D. E. Sichel, "The Resurence of Growth in the Late 1990s: Is Information Technology the Story?," *Journal of Economic Perspectives*, 14(4), Fall 2000, p. 3–22.

6 W. D. Nordhaus, "Productivity Growth and the New Economy," NBER Working Paper 8096, 2001; "Badly in Need of Repair," *The Economist*, May 2, 2002 , p. 67; R. J. Gordon, "Has the 'New Economy' Rendered the Productivity Slowdown Obsolete?," working paper, Northwestern University, June 12, 1999; "Technology After the Bubble", *McKinsey Journal*, April–May 2002, p. 1–16; Oliner and Sichel, op. cit.; J. Christopher Westland, "The Cost of Errors in Software Development: Evidence from Industry," *Journal of Systems and Software*, 62(1), 2002, p. 1–9.

7 This was pointed out nearly a decade ago in Z. Griliches, "Productivity, R&D and the Data Constraint," *American Economic Review*, 84, March 1994, p. 1–23.

8 These and other shortcomings are considered in depth in J. S. Landefeld and B. M. Raumeni, "Measuring the New Economy," *Survey of Current Business*, March 2001, p. 23–40.

9 Current approaches are based on methods developed in E. Brynjolfsson and C. Kemerer, "Network Externalities in Microcomputer Software: An Econometric Analysis of the Spreadsheet Market," *Management Science*, 42, 1996, p. 1627–47; and in N. Gandal, "Hedonic Price Indexes for Spreadsheets and an Empirical Test for Network Externalities," *Rand Journal of Economics*, 25, Spring 1994, p. 164–70. Details of current approaches are surveyed in B. T. Grimm, B. R. Moulton, and D. B. Wasshousen, *Information Processing Equipment and Software in the National Accounts* (Washington, D.C.:, Bureau of Economic Analysis, 2002).

[10] These issues are discussed at length in Chapter 5 on ownership and control of assets.

[11] A. Smith, *The Wealth of Nations* (reissued by Modern Library: New York, Jan 1994) Chapter 4.

[12] Smith actually proposed (and confused) two different "labor theories of value." In the first, value was proportional to the amount of labor consumed in production; in the second, value was proportional to the amount of labor which might be exchanged for a good. Later writers, notably Alfred Marshall, identified the theory of value as identical to the theory of price; thus value is simply how much an object will fetch in exchange. Marshall left open many questions about the practical calculation of this "price."

[13] Statement of Financial Accounting Standards No. 86: Accounting for the Costs of Computer Software to Be Sold, Leased, or Otherwise Marketed, and Statement of Financial Accounting Standards No. 141: Goodwill and Other Intangible Assets.

[14] Examples are described in K. L. Keller, "The Brand Report Card," *Harvard Business Review*, January–February 2000, reprint 00104, and in K. S. Cravens and C. Guilding, "Strategic Brand Valuation: A Cross-functional Perspective," *Business Horizons*, July–August 1999, p. 53–62.

[15] Tom Stewart, *Intellectual Capital: The New Wealth of Organizations*; New York: Bantan Books, 1998.

CHAPTER 4

Value Maps

So Geographers in Afric-Maps
With Savage-Pictures fill their Gaps;
And o'er unhabitable Downs
Place Elephants for want of Towns.

– *Jonathan Swift, On Poetry*

BANISHING THE ELEPHANTS

Through much of history, the central part of a geographical map represented the known and from there faded outward into the unknown. This phenomenon of decreasing knowledge and accuracy was doubtless responsible for the cartographer's tendency to fill unknown spaces with imaginary creatures, a feature which Jonathan Swift derides in his verse. Until recent years, geographical maps could be trusted only to report honestly the cartographer's immediate surroundings. Up to the 20th century, descriptions of distant lands were often exercises in conjecture based on inadequate surveying, theological dogma, and a lively imagination.

Elephants have vanished from modern geographical maps; in fact, today we would be most likely to distrust a map embellished with elephants, imaginary creatures, and regional princes. These represent antiquated mapmaking from a less sophisticated, technologically backward time (at least, cartographically speaking).

As cartographers' maps reflect geographical reality, so do financial accounts, analysts' valuations, credit ratings, and other statements provide maps of fiscal reality, of use to banks, investors, and other parties. No map is entirely trustworthy; maps, by their very nature, are selective representations of reality. But financial maps are still burdened with their ancient vision of reality, a perspective we owe to the Venetian friar, Luca Pacioli.

Like medieval maps, Pacioli's accounting worked well within the limits of a person's direct experience, but financial knowledge and accuracy fade outward into the unknown as we travel further forward in time, or further away from core activities in Pacioli's value map.

Unlike cartographers, accountants still embellish their "value" maps with metaphorical "elephants" – accounts, notes, and summaries that suggest complete knowledge, but really just fill in the blank spaces. Why else would "complete" and "accurate" financial statements account for only around 20% of the value of the S&P Industrials?

Pacioli's accounting was well suited to the needs of the Venetian merchant marine. It tracked "wealth" associated with consignment goods which were either agricultural, or outputs of simple cottage industries. Transactions were anything that increased or decreased the amount of silver and gold the merchants possessed – typically sales, purchases, or other transactions that reflected movement of goods in or out of the firm. Internal management was not accounted for, nor were industry and economy-wide trends, even though these might overwhelm the impact of transactions. Accounting was restricted to a tabulation of events at the boundary of the firm – what today we call "arms-length" transactions. The territory beyond is either marked terra incognita or, worse, is embellished with fanciful assumptions.

The problem arises from an accounting principle called the "entity concept" – the concept that financial accounting and reporting relate to the activities of a specific business entity and not to the activities of the owners, managers, suppliers, consumers, or regulators of that entity. This map of the firm made sense in a time that knew only proprietor-owned businesses based on trade in tangible goods with clear property rights. Several developments by the 21st century had hopelessly antiquated this vision of the firm.

Foremost is the split between owners (concerned with maximizing share price), creditors (concerned with downside risk), and managers (concerned with their jobs) in publicly held corporations, with their divergent sets of incentives.

Second, the increasing efficiency and reach of markets undermines the rationale for vertical integration, but increases the influence of suppliers and customers on the firm. Third, the shift to information goods creates ongoing opportunities for competitors to enter a market (as well as a shifting set of barriers to entry), while creating opportunities for alliances. Finally, accounting measurements that were never intended for anything but stewardship of goods on ships are being used to track all of these phenomena.

None of these developments in the firm are new. Michael Porter[1] summarized them in his "five forces": competitive rivalry, buyer and supplier influence, barriers to entry, and substitutes. They lie at the core of strategic management as espoused by authorities such as Igor Ansoff[2] and Michael Porter – the deliberate calculation and analysis designed to maximize long-term advantage. But the calculations required of management have grown significantly more complex over the past decade, and managers must be much smarter to do them well. They also need new tools. A varied array of business models have appeared that extract enormous value from new technology. In the context of these new models, though, classical tools of valuation – double-entry bookkeeping and discounted cash flow – rely on business models from a simpler era and produce results that are too often inaccurate and irrelevant.

We owe ourselves a better technology for laying out the firm's value map. In Financial Dynamics, we "map out" the processes that generate value through a Financial Dynamics innovation – the value map.

THE FINANCIAL DYNAMICS VALUE MAP

Edmond Halley, patron of Newton, student of magnetism, Latin poet and cartographer, is best remembered today as the 17th-century astronomer who predicted the periodic return of Halley's Comet. But his greatest contribution to science was the introduction of the first "thematic map" – a map that illustrates a set of facts and figures in a geographic format. Haley's map was the culmination of his two-year voyage as captain of the Paramour mapping the variations in the magnetic compass from true north. Halley's map ushered in the very modern notion that anything that can be measured – abstract or not – can be mapped.

The context of Financial Dynamics is set by the firm's value proposition (Fig. 4.1), which ultimately drives its business model as well as strategy and technology choice.

Figure 4.1 Value proposition and technology choice

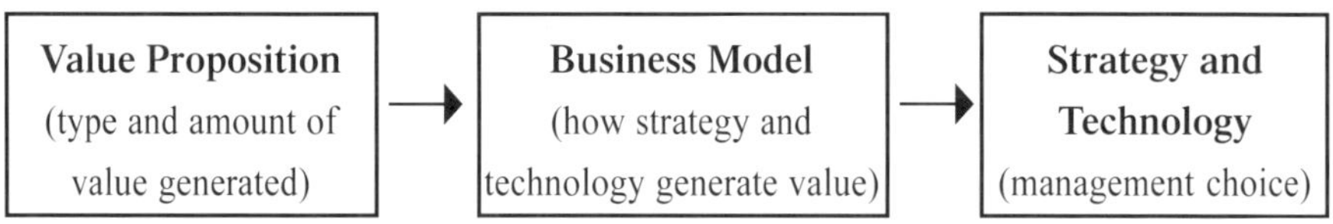

The value proposition is the business' raison d'être – the justification for committing people and resources to the operations. The ecology of the market may allow different business models to find their own niche within the competitive environment. Swatch makes low-cost watches whose styles are updated weekly; Phillip Patek makes expensive watches which the company claims are "timeless." Each can be considered different solutions to making a profit in the consumer watch market. Each demands different technology and strategy to generate value. Yet, the direction of influence is the same in both businesses: only once the idea or objective of the business – that is, the value proposition – has been articulated can management go about designing the business processes. Once business processes have been settled upon, management can make strategic or technology choices, though sometimes the processes will be selected to take advantage of a firm's particular strategic or technological advantages. The approach is Platonic: the value proposition has a permanency that is wanting in the ephemeral worlds of business processes, strategy, and technology.

Our map of value continues with a diagram of major parties that own, have claims to, or influence over the value created in a firm. These are the "principalities" of our map, and "value flows" through rivers traversing the map. The major parties with claims to value are owners (concerned with maximizing share price), creditors (concerned with downside risk), managers (concerned with their jobs), competitors (concerned with market share), and buyers and suppliers (concerned with price and timing).

Value today is seldom localized in one setting or owner, contrary to the medieval notion of value. Value in the knowledge economy arises from "ideas in action." Knowledge assets, themselves, are little more than curiosities until they are put to work. In the traditional "stocks and flows" perspective of corporate systems, stocks have played a central role in corporate systems. In Financial Dynamics, stocks, if they exist at all, are just another cost of doing business. And less is more.

The core concept in Financial Dynamics is transaction flows in the value map. "Transactions" are economic events arising either outside or from within the firm, that either consume or generate value. In traditional accounting, journal entries capture the transactions specified on the chart of accounts, and all value is measured in the cash equivalents exchanged in arm's-length transactions. The value map, in turn, consists of the key value-generating processes in which the firm has invested; and the transactions running through them.

There are also "inductive processes" that consume resources and modulate, in one way or another, the operation of key processes. These may support key processes, as is the case with the traditional "cost centers" of the firm – information processing, legal, accounting, administration, and so forth. They may be internal or external to the firm, and are often the subject of "outsourcing" decisions because of their non-central character. Geographical scaling, organizational scaling (that is, reengineering), technology acceleration, and network externalities can also be modeled as "inductive activities."

Where traditional GAAP fails, and along with it various dependent valuation methodologies such as discounted cash flow, residual income, and real options, is in their inability to formally incorporate the value map into their valuations. Because of this, it is difficult to alter modeling parameters (for example, for sensitivity and "what if" analysis) because the structure of the value map may create complex, dependent, and non-linear scaling in any of their supporting GAAP entries.

The value map uses the mathematically rigorous methodology provided by dataflow diagrams[3] to capture the structure of value, and of the competitive and strategic environment in which it functions. Industry, market, and strategic information about price and market targets can be used to set initial parameter values, which will then be accurately updated by any historical data at hand – for example, in the historical financial statements, internal documentation, or detailed transaction records.

Financial Dynamics' value maps (as shown in Fig. 4.2) are, a straightforward way of identifying the deep structure of wealth-creation in the firm. This deep structure consists of core processes that contribute to operating income; and of identifying supporting processes that are the subject of discretionary costs, and which influence the firm's ability to generate operating income. Obviously, there will be other "extraordinary items" and "claims" such as taxes which influence overall profitability. These are peripheral to the value map modeling, and are incorporated through the "static" constructs of assets, options, and claims in the Financial Dynamics models.

Figure 4.2 An example of the value map in financial dynamics

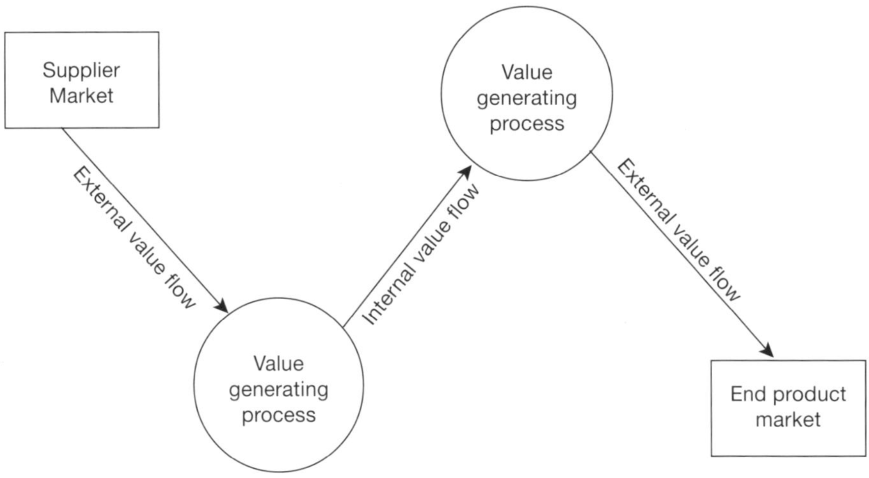

FUNDAMENTAL MODELING UNIT

The fundamental modeling units in Financial Dynamics are the *environment box*, the *process bubble* and the *value flow arrow*. Value is presumed to flow, attached to discrete quanta of value called transactions. Financial Dynamics allows these flows to be evaluated either as discrete or continuous flows; whichever is most convenient for the decision at hand. The Financial Dynamics diagram is always drawn so that there is one value flow associated with one "owner" of the process generating that value flow. In general, the entire value map is described in units corresponding to those on the rightmost value flow (that is, the final output product), and the remaining value flows are measured in terms of "equivalent units" of the final output. Value flows may be complementary, aggregating subcomponents; or they may be supplementary, adding value to a product in stages (Figs. 4.3 and 4.4).

Figure 4.3 Complementary processes

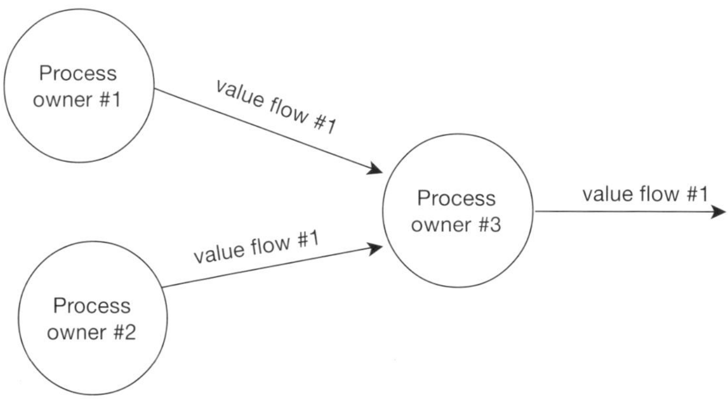

Figure 4.4 Supplementary processes

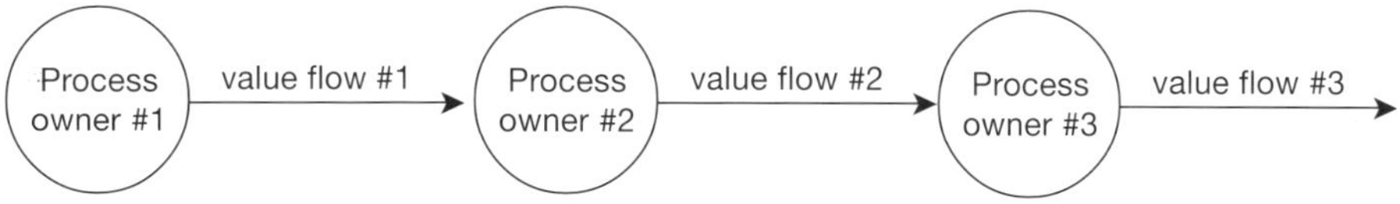

THE FUNDAMENTAL "QUANTUM" OF VALUE

During the last two decades of the 19th century, colonialism had caused cartographers to reflect on the shortcomings of their discipline in its ability to generate a world map. By late 1884, less than one-ninth of the land surface of the globe had been surveyed or was being surveyed.[4] This patchwork of surveying had been confounded by a variety of standards of measurement. Each country had gone its own way, developing its own traditions from a chauvinistic perspective.

Without standardization, mapmaking could never provide an unambiguous language. England had a semblance of standards: Henry I decreed that the yard was the distance from the tip of his nose to the end of his thumb; the length of a furlong (furrow-long) was established by the Tudors as 220 yards, and Elizabeth I declared a mile to be eight furlongs (increasing the Roman mile of 5,000 feet to the modern mile of 5,280 feet). France was in much worse shape, having different standards from province to province: the most commonly used measure on maps was the toise, which equaled six French feet (which were slightly longer than English feet) and was derived from half the width of the main gate of the Louvre. Yet, it was ultimately the French Academy of Sciences that developed the "metric" standard (from the Greek metron = "a measure") that prevailed: a meter was one ten-millionth the meridian distance between the Pole and the Equator; a gram was the mass of a cubic centimeter of water; a liter was the volume of a cubic decimeter. The immediate effect of the metric system on mapmaking was to introduce a simple, universal language for expressing map scale. In October 1884, the U.S. government invited 25 nations to the first International Meridian Conference, in which France and England agreed that the prime meridian would run through the Observatory at Greenwich, England, in exchange for the British adopting the metric system. Only then did mapmaking embrace a universal and unambiguous standard of measurements necessary for mapping the world.

Unfortunately, financial accounting still operates with a patchwork of metrics developed by each firm or discipline going its own way, developing its own traditions from a chauvinistic perspective. In truth, much of what passes as business expertise is simply wasteful effort translating one local dialect to another. The commonly adopted "standard" of measure is the U.S. dollar (see Fig. 4.5 for an idea of how "standard" this really is). Consider how this measure changes from region to region, or from business to business, commodity to commodity.

Figure 4.5 The value of a dollar

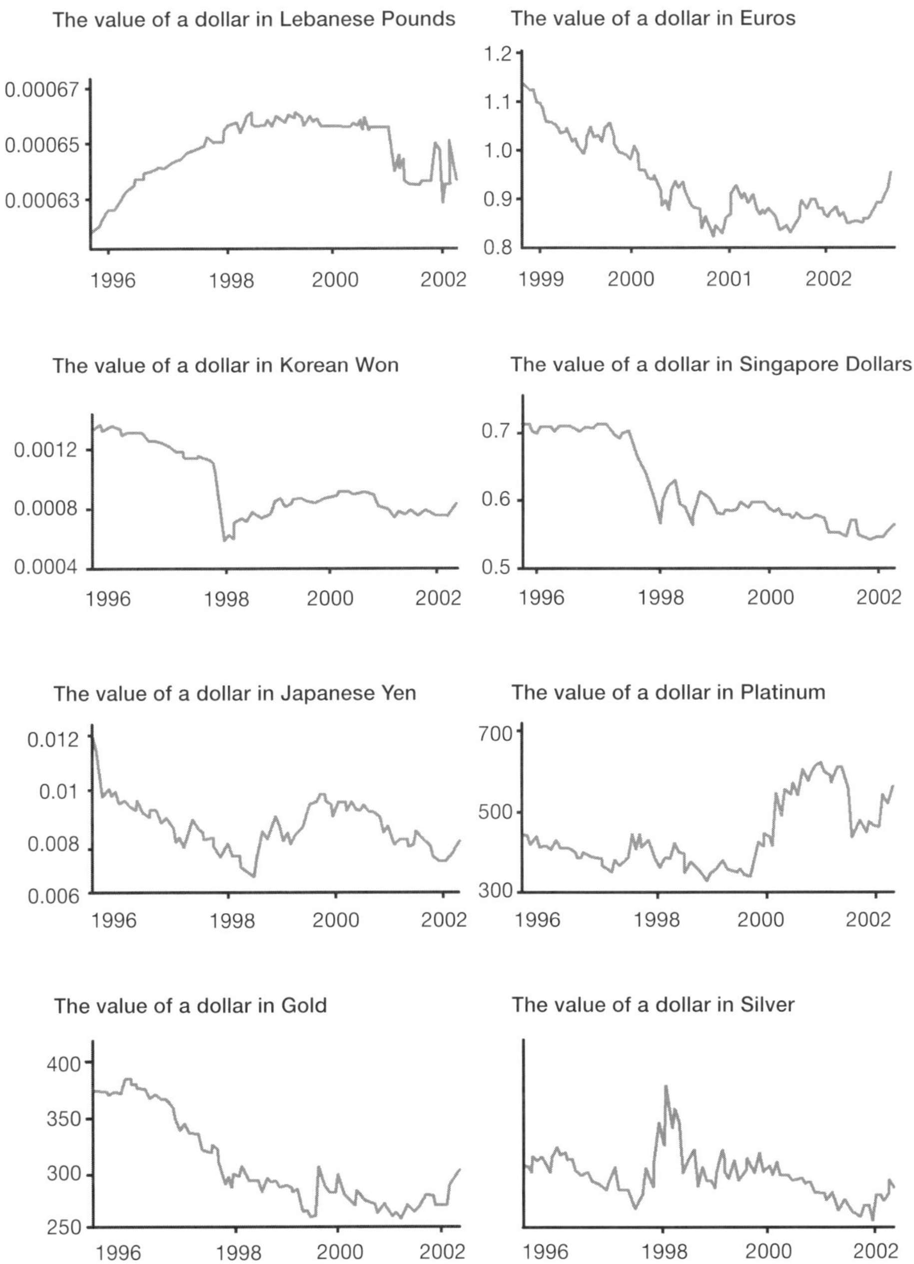

Obviously, the dollar is a measurement unit that varies from region to region. True, there are exchange rates that price out these differences (just as there were conversion tables from French toise to English yards in 19th-century mapmaking). But any particular business operates in a specific region, purchasing particular commodities, selling its products in exchange for specific currencies. The knowledge economy clearly exacerbates problems with a monetary denomination. Many knowledge products have no market, nor is there a clear way to value them (for example, consider a patent or copyright).

In the environment of the knowledge economy, then, how can you distinguish good management, good business models, and good accounting from simple good luck? This was a major problem in the 1990s' dot-com bubble. It was difficult to know what a "good" business model was, because so many of the models were novel and untested. Money and effort chased after business models that didn't make sense, but were – for a while – very lucky. And when their luck ran out, managers in some firms (e.g., Enron and WorldCom) augmented the gap in their revenues with creative accounting (e.g., fraud). These sorts of problems will only grow more widespread and common as the knowledge economy grows and business becomes even more complex.

Financial Dynamics takes a non-denominational and opportunistic approach. We allow a customized definition of the basic transaction across the value map as an alternative to expressing all physical and knowledge assets first in dollar terms, and then aggregating them (a process that typically leads to large errors in valuation). This avoids problems with monetary units which arise in four areas:

1. *Inflation / deflation:* They are unstable over time.
2. *Erratic exchange rates:* They are unstable across global operations.
3. *Market failures:* They fail to properly capture value where markets are either thinly traded or nonexistent.
4. *Rapid depreciation:* They fail to properly to assess value where there is rapid depreciation of goods – this is especially a problem where technology is accelerating rapidly.

Non-monetary bases for valuation, as well as a graphic map of processes, have been used in several valuation approaches over the past half-century. The cost accounting techniques that arose around Fredrick

"Speedy" Taylor's factory accounting and efficiency movements after World War I provided an equivalent units of production (EUP) technique for putting all processes throughout the factory on a common metric – essentially, units of finished and salable product.

Activity based costing (ABC), which was popular in the 1990s, extended cost accounting concepts to a broader base (including a broader scope of non-monetary bases or "activity bases") than cost accounting's "finished goods." ABC laid out the organizational ownership and control structure in a value map, similar to the value map prescribed in this chapter.[5] Unfortunately, ABC has languished in recent years. Its reputation was sullied by association with the Business Process Reengineering fad of the early 1990s, where it was used to justify change and lay-offs. More subversive, though, may have been the general failure of ABC's proponents to clearly define activity bases in a fashion that allowed easy business application.

On a much grander scale, the Soviet Union's Soviet State Planning Committee[6] ran the national economic policy for nearly seven decades on a non-monetary basis. In 1919, Lenin debauched the Russian currency by cranking up the printing presses (a stratagem used repeatedly throughout Russia's history, most recently in 1997). His goal was State Capital expropriation of all of the major industries in pursuit of the "commanding heights" of the Russian economy. Once money was superfluous, Russian production quotas were set by the state through a series of five-year plans. By design, supply and demand always adjusted to meet these quotas (surprise!). Money became redundant, and all accounting up and down the supply map was done in units comparable to the cost accountants' equivalent units of production (EUP). Under free markets, such a system is still useful to planners and managers, even if it does do a terrible job of defining the schedules and preferences of consumers and producers.

The legacy has been a modern Russian industrial complex that conducts 40% to 70% of its internal trade via barter. A barter economy must base value on *ad hoc* assessments for each exchange, which potentially requires enormous information processing overhead when transaction volumes become large. The modern solution is to set reference prices which can be used to determine transaction ratios.

MODELING GRANULARITY AND TRANSACTION SIZE

Let's get back to our original issue of a customized definition of transactions in terms of physical- and knowledge-based units, as opposed to a dollar basis. Financial Dynamics solves this problem by forcing the analyst to choose a fundamental metric of value that is unique for a value map. This unit may then be converted to other units, but only if the method is transparent. This is like the equivalent units of production approach used in cost accounting including its recent extensions for activity-based costing.

Costs and revenues are both lumpy – and usually they come in lumps of significantly different size. Computer scientists refer to the "granularity" of something when trying to articulate the concept of "lumpiness" in an abstraction (like money). We will adopt the computer scientist's terminology here, which is able to, in a rather sloppy way, indicate the size of the lumps as well.

Consider an example from the software industry. The fundamental unit by which a software business is measured could be perceived from the supply side or demand side. On the supply side, the unit of work is the packaged software product. This will take many man-years of effort to produce, but once produced, the firm can make an unlimited number of copies to sell. Thus, the value is the full cost of production, less the projected life-cycle revenues, less the projected life-cycle support for each software package produced. From the demand side, the logical unit is a single copy, the unit used for measuring sales. From a value modeling standpoint, adopting the "quantum" to be the packaged software product makes more sense than a single copy of the software. Management's strategy and policy are applied uniformly across all sales, but are unique to the project. Thus, the appropriate level of granularity for "quanta" or "transactions" is at the software project level. All measurements throughout the value map should be expressed in terms of unique software products (whether they be versions of a given product, or customized installations). The knowledge economy increasingly allows costless copying – today, of pure knowledge such as software, films, and music; but tomorrow, this may extend to electronics and fashion through flexible manufacturing and, ultimately, the nanabox. In this case it is the *process* – with its set-up costs, learning curve, and sunk costs – that is the fundamental quantum of value.

This fundamental quantum of value is called, in Financial Dynamics, the "transaction" – an economic event affecting the value of the firm. Each transaction is characterized by two measurements: (1) a value; and (2) a point in time at which the value-generating process generates the transaction.[7]

The value chain is a conceptual model of the value transformation in firms which was introduced by Michael Porter in his 1985 book *Competitive Advantage*.[8] Porter's idea derived from the economic intuition of production that every firm is a collection of activities that are performed in order to design, produce, market, deliver, and support its product. All these activities can be mapped using a value map. A firm's value map and the way that it performs value-adding activities reflect its history, its strategy, and the underlying economics in its markets.

In Porter's view, there were eight sorts of activities in a typical value map. There were five types of primary activities: inbound logistics, operations, outbound logistics, marketing/sales, and service; and four types of supporting activities: procurement, R&D, human resources, and infrastructure. *Competitive Advantage* was written in the early 1980s; the personal computer had just been introduced, and information technology was limited mainly to back-office accounting systems. The modern pantheon of corporate activities – especially knowledge-intensive ones – has evolved substantially since then. Today's value map must map a rich, complex, and inherently knowledge-based set of primary and supporting activities. The rate at which new technology adds competitive niches means that any definition must be open-ended; an entirely different environment than the one Porter faced two decades ago.

In knowledge-intensive businesses, an updated perspective may be more suitable. In this perspective, transaction flows circulate around the core processes in the firm's value map. Core processes directly involve the creation and delivery of a product or service. In Financial Dynamics, we define core processes as those essential to value creation. Supporting processes add cost, and ostensibly make core processes more efficient. There are usually many different options for supporting core activities – for example, make or buy, vertical integration, outsourcing, and so forth. As a consequence, the inherent cost of the latter activities will often be difficult to assess. Generally, though, they may be considered to fall into four categories:

- transportation, coordination and communication activities;
- administrative activities (administrative systems are support activities that improve the efficiency of core activities):
- quality control activities; and
- activities which promote process efficiency.

Porter's value chain perspective has proved popular, and consultants and academics have over the years incorporated it into varying formal methodologies – from extensions to XML called Value Chain Markup Language, to extensions of Michael Hammer's Semantic Modeling, which were the basis of the reengineering vogue a decade ago.

Value-generating processes on the value map transform the transactions to generate value for the firm. This value is assessed by the ultimate consumer of the product or service, and – from the firm's point of view – can be assumed to be the aggregate of revenues, costs (both capital assets and expenses), and adjustments to risk (for example, hedges such as are common in commodity-intensive businesses).

Such an approach makes it much harder to hide wealth, costs, and other relevant influences on value. Rather than periodically inventorying asset and liability balances, Financial Dynamics tracks the value flows that change these inventories, and considers the variance of these flows. Consider the tactics used by telecommunications company Qwest to inflate income through accounting for offsetting sales and purchases of bandwidth. In the first six months of 2001, Qwest sold US$857 million worth of bandwidth to other telecom companies which was accounted for as revenue; at the same time, it also bought US$450 million from some of those same companies, which was accounted for as an asset. This accounting fiddle pushed Qwest's revenue growth for the six months from 7.5% to 12.0% at a time when insiders were selling off substantial stakes in the company at inflated stock prices.[9] A purely transaction flow accounting – as, for example, would exist in a Financial Dynamics valuation – would not allow this sort of manipulation, since the artificial distinction between "real" and "nominal" accounts is discarded in Financial Dynamics.

Recall that the artificial distinction between "real" and "nominal" accounts is a direct result of the combination of double-entry bookkeeping and the stored wealth perspective of GAAP accounting. Double-entry bookkeeping records every transaction twice – once as a debit and once as a credit – originally to enforce a crude sort of error control long before there were computers or calculating machines. The stored wealth perspective arose at a time when accounting involved accumulating individual transactions into aggregates, with the objective of calculating the value of physical assets owned. These conventions predated the existence of futures contracts, depreciation, information assets, risk hedges (for example, insurance, options, etc.), computers, the Internet, or almost any other major factor in the modern world. Taken together, these conventions required two types of accounts: the "real" accounts of the income statement, which tabulated the aggregate totals of transaction flows; and the "nominal" accounts of the balance sheet, which tabulated the net effect of inflows and outflows of physical assets (via transaction flows) on physical stockpiles and obligations to creditors. Obviously, this does not paint a particularly accurate picture of business in the 21st century.

VALUE FLOWS

At the core of Financial Dynamics modeling are value flows. Their representation is consequently the single most complex aspect of this valuation approach. In exchange for added complexity, the method (1) allows for creation of independent models of value as well as volume of transactions (2) incorporates specific strategy and technology models, i.e., the explicit or implicit models underlying management's strategic choice and the consequent future value creation, and (3) provides a statistically valid alternative to ad hoc growth models implicit in traditional valuations.

Financial Dynamics refines the validity of corporate models by recognizing that value is delivered in discrete quanta called transactions. *Transaction price* measures the value delivered; *transaction volume* measures the number of transactions processed in a given time period (the default accounting time unit is one year); value flow is total value generated in a given time period. In practice, the distributions of firm account balances tend to be highly kurtotic (fat-tailed), and right-skewed (because negative values are prohibited). These tendencies are the result of commingling different value-generating processes in the same account.

More importantly, particular management strategies may tend to affect one component differently than the other. Let's assume that management is assessing the merits (i.e., impact on future value flows) of a high volume / low margin product (think Saturn automobiles) or of a low volume / high margin product (think Ferrari). The first strategy raises volume while lowering unit value; the second strategy raises unit value while lowering volume. If only value, and not its components are considered, the two might well be seen as equivalent. But if instead we think in terms of volume and value components, the comparative impact of different strategies on potential losses, risk and error become explicit.

Very little research has actually been conducted on asset market volumes. Where it has been done, the influence of volume and liquidity on price volatility are most often addressed. Thus there tends to be a significant asymmetry in the amount of research done on market prices and on trading volumes, and we cannot necessarily look to that research for guidance in our current task. Fortunately, within a firm or project context, we know a good deal more about transaction volumes because they are so central to capacity planning, utilization, and discretionary budget. Managerial strategy and policy are most important in ascertaining transaction volumes; market forces are most important in ascertaining transaction prices.

GROWTH, CHANGE AND RISK

Traditional DCF analysis commonly assumes an *ad hoc* two-stage growth model for forecasting future cash flows (the so-called Gordon growth model). This assumes that there is an initial stage of rapid growth in a firm, some constant rate of growth which applies up to the forecasting horizon, which is typically set 10 to 15 years in the future. Past that horizon, convention uses two approaches to calculate a so-called 'terminal value' – (1) perpetuity, and (2) price / earnings multiple approaches. The perpetuity approach assumes that the cash flows beyond the forecast horizon are 'perpetuity' – an annuity payable indefinitely. This approach assumes – primarily for sake of argument – that the firm reaches a steady state of operations. Alternatively, if the firm's business model is stable (a big *if* in the knowledge economy) the price / earnings multiple approach assumes that the market will ultimately value the firm at some multiple of earnings. Industry averages may be used, or, for lack of better numbers, it is common practice to use a multiple in the range of 5 to 15.

Lacking information to the contrary, it seems reasonable to assume constant growth (or shrinkage, though this seems suspiciously rare in practice) based on prior trends in the objective historical data. With additional information – e.g., seasonal cycles, market trends and so forth – a wide range of econometric and time-series statistical tools have been developed in economics to improve forecast accuracy. The Financial Dynamics approach lends itself well to the incorporation of these techniques. But for the duration of this book, I will limit discussion to ordinary least squares models with a constant growth rate over the forecast period.

The Gordon growth model[10] was introduced in a popular finance text back in the 1960s. Henri Pagès (1999)[11] suggests that Gordon's growth model is inherently incompatible with non-stationary growth of firm or project value. A more glaring problem arose in during the dot-com bubble – there is little justification for the 'kink' that occurs in growth between the first and second stage. This kink often took extreme values in financial plans that appeared during the late-1990s dot-com bubble, where it was not uncommon to see 80%-120% of the firm value in the terminal valuation.

Financial Dynamics adopts a more transparent and defensible approach to growth. It discards the planning horizon altogether. Growth is seen as resulting from two sources: (1) growth induced by managerial strategy; and (2) growth induced by the external supply and product market factors – what we call the business environment.

---

1 Michael E. Porter , *Competitive Strategy* (New York: Free Press, 1998); Michael E. Porter, *Competitive Advantage* (New York: Free Press,1998); Michael E. Porter, *The Competitive Advantage of Nations* (New York: Free Press, 1998).

2 Igor Ansoff, Edward McDonnell, Linda Lindsey, and Stephen Beach, *Implanting Strategic Management*, 2nd edn (Englewood-Cliffs: Prentice Hall, 1993).

3 Dataflow diagrams were shown to have a 1:1 mapping to Petri nets (see M. Adler in *An Algebra for Data Flow Diagram Process Decomposition*, Long Beach: CA, IEEE Trans. Software Engineering, February 1988). Because of this, they provide an unequivocal model of firm operations that can be incorporated into a rigorous and scientific model of firm operations and transaction processing. Such rigor is needed if the predictions of the model are to be reliable.

4 J. N. Wilford, *The Mapmakers* (New York: Vintage Books, 2001, p. 253).

5 Financial Dynamics' value map borrows symbols and terms from dataflow diagrams (DFD), whereas ABC typically used an IDEF0 modeling tool, with more involved terms and modeling techniques. Both can be shown to be equivalent to Petri Nets, and thus provide the same language richness. DFDs are simpler and more intuitive; thus, the choice was made to pattern value maps on the DFD modeling approach. (This short section is based on *Business Process Improvement* v.1.0. Office of Information Technology (AIT), Federal Aviation Administration, Washington, DC, November 30, 1995).

6 The Soviet State Planning Committee (*Gosudarstvennyi planovyi komitet – Gosplan*) was primarily responsible for creating and monitoring five-year plans and annual plans.

7 We ultimately can make these consistent with the statements that we are used to; though not perhaps without throwing out some data. Through mathematical manipulations, these flows may be reinterpreted into those corresponding to account values appearing on GAAP income statements.

8 M. E. Porter, *Competitive Advantage: Creating and Sustaining Superior Performance* (New York: The Free Press, 1985).

[9] R. Grover, C. Palmeri, and P. Elstrom, "Qwest: The Issues go Beyond Accounting," *Business Week*, March 25, 2002, p. 66–68.

[10] M. Gordon, *The Investment, Financing and Valuation of the Corporation* (Homewood, IL: Irwin, 1962).

[11] H. Pagès, "A Note on the Gordon Growth Model with Nonstationary Dividend Growth," Bank for International Settlements, Working Paper No. 75, Basel, Switzerland, 1999.

CHAPTER 5

Ownership and Control

No man acquires property without acquiring with it
a little arithmetic also.

– Ralph Waldo Emerson[1]

What would life be without arithmetic,
but a scene of horrors.

– Sydney Smith[2]

CHARTING TERRAIN ON THE VALUE MAP

In Chapter 4, we introduced the notion of a "value map." The "bubbles" represent distinct "entities" in the value map – entities being firms, individuals, or anything else that "owns" the value-creating processes in this part of the value map. At the simplest level, each "bubble" represents the legal owner of the value-creating process; this is the perspective promoted in accounting's "entity concept" principle. But we have come to accept in the past two decades – by way of models such as Porter's five-forces model – that the most important influences on value creation can arise from channels outside of ownership, and furthermore, legal ownership does not necessarily ensure control. In this chapter, we tackle the increasingly difficult task of defining ownership of valuable things or methods.

Technically, *ownership* is the *legal relation* between an individual, group, corporation, or government, and *property*. The object may be tangible, such as a building, or completely the invention of law, such as a patent, copyright, or annuity; it may be movable, such as a car, or immovable, such as land. Ownership of property means, at a minimum, *excludability* – that is, that government helps to exclude others from the use of something you own without your consent, possibly at a price.

The English word "property" derives from the Latin *proprietas* – a word derived from *proprius*, an adjective meaning "peculiar" or "own," as opposed to *communis*, "common," or *alienus*, "another's"; thus, it came to have dual meaning as "the peculiar nature or quality of a thing" and "ownership." Before the 17th century, it was rare to find the word "property" used in its modern sense of an object of legal rights. As the need arose for a category to describe the sum of the rights, privileges, and powers that an individual could have with respect to a thing, the Romans, followed by the English, chose a noun derived from an adjective that means "own." The category at once described the concept and also the tendency. As time went on, the tendency took on an independent life. In modern legal systems, property came to represent one of the rights of the individual under a given government.

Possession and *control* are considered to be major factors in ownership – the acquisition of either a considerable degree of physical control over a physical thing, or the legal right to control intangible property, such as a credit. The concept of intangible property rights is a modern one. Civil law, following Roman, has tended to deny the possibility of legal possession of anything that cannot be touched. Abstraction of contract rights and duties did not evolve until the 19th century. The conceptual breakthrough came in negotiable instruments. Because a negotiable instrument is freely assignable, and since it is enforceable without regard to the underlying transaction, it has two of the general characteristics of property. There has been a subsequent expansion of the types of intangible rights granted by the government. The oldest of these are the exclusive rights given by the state to encourage and protect authors, inventors, manufacturers, and tradesmen. Copyright, the exclusive right to prohibit the copying of a piece of writing or a work of art or music, is almost universally regarded as a property right. Patents, the government-granted right to the exclusive use of an invention, and trademarks, the government-granted exclusive right to market one's product with a given distinctive sign or symbol indicating its source, receive similar treatment.

Concepts of property and ownership have been repeatedly challenged over the past 50 years, with profound effects on accounting. Accounting, which succumbs to the dictates of the "entity concept," is interested in the value accruing only to the legal owner of the value process. But as value-generating processes become more knowledge-intensive, ownership, in the form of complete possession and control, becomes more difficult to enforce. The knowledge components of modern products tend to be non-rival – that is, two firms can use the same knowledge without degrading its utility; in fact, utility increases because of the network effects. They also tend to be non-excludable, thus it becomes difficult to exclude firms from using knowledge.

Steven Pinker in his book *How the Mind Works*[3] argues compellingly that this non-rivalry and non-excludability of knowledge have been central to human evolution. His claim is that early humans formed into groups to defend against predators and improve their foraging. Group living set the stage for the evolution of human intelligence in two ways. With a group in place, the value of having better information is multiplied, because information can be given away and kept at the same time (it is non-rival). Smarter animals living in a group enjoy the benefit of their knowledge and the benefit of whatever they can get in trade for their knowledge. But they share the disadvantages of a madding crowd – neighbors compete over food, water, mates, and nests; social animals risk theft, cannibalism, cuckoldry, infanticide, extortion, and other treachery. In such an environment, the right evolutionary track is to get smarter. So, the largest-brained and smartest-behaving species are social: bees, parrots, dolphins, elephants, wolves, and sea lions, to name a few. Humans get smarter by extending their brain development for a year after birth, resulting in a huge organ adept at inference and abstract thought. (If our body development were similarly extended, we would be 10 feet tall and weigh half a ton.) Information technology and the knowledge economy are modern aspects of the continuing evolution of human intelligence. But I digress.

Back in the 1950s, companies prided themselves on the conservatism of their accounting. What changed their mind? One factor was the rise of inflation, which widened the gap between a company's book value on its balance sheet and its market value. Companies could book a profit simply by selling an asset at the higher current price and pocketing the difference over what they had paid for it. A second factor was the emergence of new sorts of assets and liabilities. A third was the evolution of the modern firm and the blurring of corporate boundaries. Joint ventures, leasing, and special-purpose entities all made it harder to determine whether a particular activity should be consolidated into a company's accounts.

The growing use of derivatives and off-balance-sheet financing and the rising importance of intangible assets such as brands and goodwill have all posed challenges to traditional accounting, none of which has been resolved entirely satisfactorily. And serious distortions have been introduced because share options awarded to employees, clearly an employment cost, are not charged against profits. The U.S. Federal Reserve Bank reckons that this caused profits to be overstated by an average of 2.5 percentage points a year over the 1996 to 2001 period.

OWNERS AND STAKEHOLDERS

Ownership of a given value-generating process is important in Financial Dynamics, as it dictates those parties that will receive, and how much each will receive, from a particular set of future value flows. It also should be separated from notions of control over strategy and policy, or historical control. We mentioned earlier that the economist William Baumol[4] estimated that, on average, less than 20% of the total value from innovations goes to those who invest directly or indirectly in that innovation. The remainder of the benefits spill over to consumers and society at large, in higher quality or utility of products, lower cost, or lowering undesirable side-effects such as pollution.

Allow me to step out on a limb for a moment to suggest that, on average, one can infer from Baumol's numbers that a typical technology-intensive value process might yield returns of 20% to the managers and legal owners (that is, stockholders) of the process; 40% to the customers of the product through reduced cost and higher quality; and 40% to society at large through taxes, regulation, and social policy. This would imply that technology-intensive processes need to be assessed from a viewpoint akin to the "*stakeholder*" perspective; *not* as a matter of corporate altruism but, rather, as a result of the non-rival (that is, my use doesn't deprive you of using information) and non-excludable (that is, I can't exclude you from access to information) character of knowledge-intensive products and processes. Since much of the content of a new piece of software, a new telephone, or whatever, is in the design and technology, this can easily leak off to rivals and be commoditized, leading to lower customer costs and higher quality. The ease with which knowledge-intensive processes can be reengineered – for example, by reprogramming their software – and the latitude that governments have to dictate engineering in products such as cars, telephones, and so forth, allows government regulators to extract further value at the stockholders' expense.

This 20-40-40 rule is no more than a 'rule of thumb.'[5] But without evidence to the contrary, and in a competitive market, it is very unlikely that 100% of value will be retained by the legal owners of the process.

OWNERSHIP IN A KNOWLEDGE-INTENSIVE BUSINESS

In my previous book *Valuing Technology*, I noted that I was unable to offer an exact system for computing the value of knowledge-intensive companies. In this book, I have set as my goal the construction of a comprehensive system of valuation whose core methodologies address the unique challenges of technology and knowledge-intensive business. This chapter lays out the infrastructure for such a method.

Financial Dynamics serves several subsidiary goals, all of which have been broached by this point in the book. I want to provide tools to move strategy – management's plan for the way that ideas and technology are put into action – from an *ad hoc* art form to an objective, quantifiable, and scientific discipline based on comparative value. The knowledge economy presents investors and managers with unique challenges, because the accounting metrics that we have traditionally used to assess performance are increasingly divorced from fiscal reality.

Knowledge-intensive companies are likely to produce things whose market value – individually or collectively – is only loosely connected to the value of the materials, labor, and capital input. Thus, valuing the bundle of processes and property rights that make up a corporation is difficult through traditional double-entry accounting. Financial Dynamics adopts a dynamic, process-based, value flow orientation and turns away from the classical stored wealth perspective. This allows a comprehensive, accurate, and objective basis for forecasting future earnings and cash flows. It fits better with modern concepts of present value and discounting over time to obtain firm, project, and asset valuation. It should also improve the accuracy of older approaches such as discounted cash flow, which try to artificially graft concepts of intertemporal economies, discounting, and contingent commodities on to a centuries-old framework of stored wealth. I will also argue why Financial Dynamics will better measure, classify, and report the real activities of knowledge-intensive business.

In the vernacular of financial markets, the Financial Dynamics methodology provides "fundamental," as opposed to "technical," analysis. Financial Dynamics focuses on firm operations, and forecasts them into the future based on specific assumptions about managerial strategy.

THE VALUE INFRASTRUCTURE

Individuals are the producers of valuable things, as well as their ultimate consumers. They define a thing's value. But it is different groups of individuals that are responsible for the various parts of the value infrastructure. Financial Dynamics perceives each firm as a bundle of value-generating processes, each centered on a particular class of product or service. Specifically, there are three domains in which value is generated – operations, the environment presented by strategy, technology and market choices, and the capital markets (Fig. 5.1). The managers of the firm control the *business operations*. Their feasible set of value-generating activities is, in turn, subject to the sum total of activities in supplier and consumer markets, as well as available technology; managerial *strategy* is the vehicle that aligns business operations with its environment. *Capital markets* provide financing for business operations. Idiosyncrasies in the way particular markets acquire and process information about business operations and strategy can result in assessments of firm value with variable accuracy.

Figure 5.1 The Three Value-generation Domains

Financial Dynamics provides a language for the delineation of *business operations*, as they currently exist in the firm, and as they are envisioned to evolve in the future under management's strategy. The language provides a synopsis of the *product and supply markets* along with *technology* and the corresponding managerial *strategy* that integrates them into operations. The language generally ignores *capital markets* because the value that capital markets add to the *fundamental* valuation of the firm can be highly idiosyncratic, and subject to different forces than fundamental valuations. For example, studies of the capital asset pricing model (CAPM) suggest that around one-third of the price movement in stock markets is driven by market-wide behavior, with very significant feedback effects. The *technical* analysis of capital market behavior has been studied extensively in the finance literature, a fact that can be applied to extend the Financial Dynamics model to include the valuation of financial instruments, as well as the firm value.[6]

MAPPING THE BUSINESS MODEL

Prior to the "reengineering" vogue of the early 1990s, financial analysts tended to focus on the product or service produced by the firm. But reengineering is about creating value by restructuring the way the firm does business. Its value proposition was based on changing the business model of the firm. By the mid-1990s, newly formed dot-coms and acquisitive giants alike were touting their "business models." The finance community was caught off-guard, and consequently generated a variety of faulty analyses that claimed to capture the "true value" of the business model of firms that had neither assets nor a history of earnings. In retrospect, much of this was silly, if not fraudulent – but it was received wisdom at the time. And the accounting and finance communities have yet to explain their failed valuations.

Financial Dynamics addresses the need to reconcile strategy and business models with value generation. It defines roles for (1) the strategic business model, (2) technology change, and (3) risk and uncertainty. The latter is essential, yet risk is an area in which financial accounting is mute. Accounting was designed in the good old days, when you were certain that the fertilizer you paid for in an arm's-length transaction would retain its value to the end of the planting season – and thus should be kept on the books as an asset at historical cost. Risk, uncertainty, and market volatility are inherent aspects of doing business today – they should appear explicitly in financial valuations.

To do this, Financial Dynamics synthesizes techniques from three areas: (1) managerial strategy; (2) statistical inference; and (3) flow schematics and models which evolved in engineering. Our objective is to improve the accuracy, informativeness, and risk assessment of valuation models – a rubric that my arguments to this point should have convinced you includes financial accounting, as well as techniques such as discounted cash flow, residual income, and other models from the investment community.

The fundamental quantum of value in financial dynamics is the "transaction" – an economic event affecting the value of the firm. Each transaction is characterized by two measurements: (1) a value; and (2) a point in time at which the value-generating process generates the transaction. A sequence of value flows – perhaps each managed or owned by different groups – comprises a "value map."[7]

Figure 5.2 might provide a typical value map for a consumer product such as an automobile.

Figure 5.2 Example of a value map

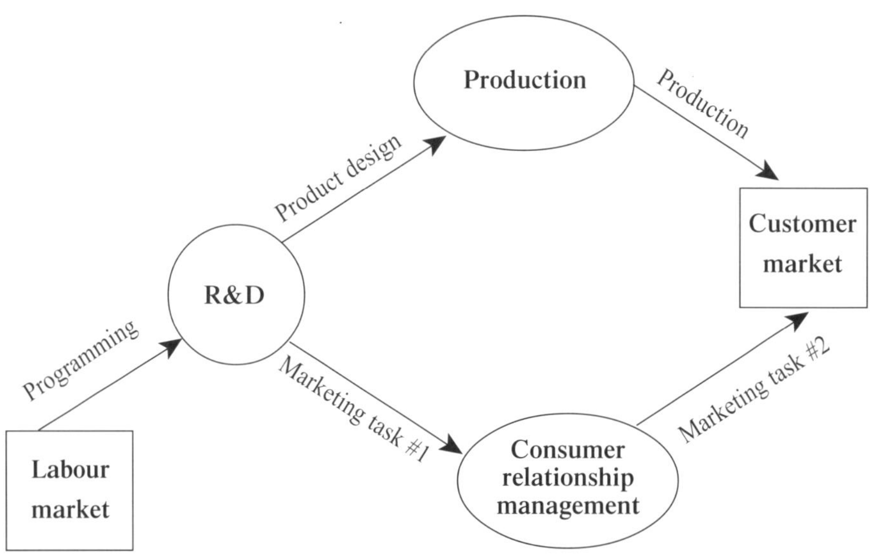

The value map consists of three types of symbols: (1) the value flow (arrow), which is the main focus of the Financial Dynamics modeling; (2) the manager/owner/controller (bubble) of the outflowing value flow process; and (3) the external market (box), which marks the boundaries of the value map and constraints on the value generating processes.

Financial Dynamics is process-centric. It ignores a firm's stockpiles of tangible, physical assets in valuing the company. If indeed these have a proven market value (most often in the case of a widely traded and non-perishable commodity), the stockpile's value can easily be added to the Financial Dynamics value. Otherwise, tangible assets are considered highly perishable – due either to technological obsolescence or to physical deterioration – and are treated as sunk costs. This is my side of the dialectic, designed to emphasize that in the knowledge economy, all that matters are ideas, strategy, processes, and the way they unfold into the future; tangible "things" are just a cost of doing business. This is especially apparent in developed economies, where the percentage of GDP generated by the production of tangible goods has fallen to around 20%. (Not surprisingly, the percentage of firm value accounted for by GAAP has also fallen to around 20%.)

In being process-centric, Financial Dynamics also discards any illusions of value being generated solely by owning a patent, copyright, brand, or other chunk of intellectual property. These, again, are only costs of doing business. Unless ideas are put in motion; unless they are formally integrated into a value-producing process, they need to be treated as sunk costs and assets with a value of zero. Part of the value-generating process must involve controls over security (for example, of trade secrets) and quality (for example, of brands).[8] The non-rival nature of intellectual property allows ideas generating competitive advantage to be quickly integrated into a competitor's product, possibly benefiting the customer, but certainly not your own profitability. And brands, unless jealously guarded and maintained, can turn quickly from assets to liabilities. In recent history, two brands – Firestone and Arthur Andersen – turned from industry leaders to pariahs overnight. Their firms did not significantly change, but perceptions did. The end result was a dramatic reversal of fortunes.

VALUE FLOWS ("ARROW")

In managing a particular value flow or product line, managers confront two classes of decisions – how much to produce, and at what price to sell. In recent years, the systems to support these two classes of decisions have been widely automated, using supply-chain and customer relationship management software. For example, retailers rely extensively on barcode scanners with the potential to tell precisely how well products are selling – sales compared with prior months, mark-ups, how much is in stock or in transit. In progressive retailers such as Wal-Mart, the systems are so developed that decision-making can be made at any level – any Wal-Mart store worker has the power to lower the price on a product if he spots it cheaper elsewhere. Such capability doesn't come cheaply – the computer controlling Wal-Mart's logistics is the world's most powerful after the Pentagon's. This level of technology investment is necessary to coordinate its 1.2 million employees (Wal-Mart is the world's largest private-sector employer) and sales (it has the second-largest sales revenue in the world, after ExxonMobile, accounting for 60% of US retail sales and 8% of total consumer spending exclusive of cars and white goods). The ability to segregate value and volume decisions reflects the fact that they may often be made in different parts of the organization – store workers decide price, whereas truck drivers and purchasing determine volume.

Financial Dynamics recognizes the importance of these two classes of decisions, and deliberately segregates the analysis of transaction *quantity* from transaction *value.* The number of transactions per period is a flow rate, while the value of each transaction is a rate per transaction. Value in the Financial Dynamics sense is a broader term than "price" and refers to the value added by each transaction – the marginal operating income of a transaction. Both of these are random variables – they take on specific values over a range described by their probability distribution. In a competitive market, the transaction volume and price define the demand curve; an analogous concept can be applied to specific value flows on the value map, though since the assignment of value is intrinsic rather than being set by the market, there is little insight to be gained from it.

Value flows display the value-generating processes of the value map. Value flows consist of transactions, each contributing a value. Financial dynamics describes the *statistical distribution* of *individual transaction values* and of *transaction quantities*. The *product* of *value* and *quantity per time period* is the *value flow per time period*.

The methodology is flexible in its use of the term "value," so long as that use is consistent throughout the prediction process. I use the term "value flow" to encompass traditional "cash flow" (which suffers from ambiguous definition in practice). But almost *no* knowledge economy companies would count their wealth in terms of inventories of stored assets. Management is probably more concerned with future revenues, or securities market value. This choice is left to the analyst, and indeed is likely to vary from firm to firm, or from analysis to analysis, depending on the use of the valuation.

PROCESSES ("BUBBLE")

Ownership of wealth is depicted by a process bubble. A particular group of investors, managers, or stakeholders is presumed to hold ultimate responsibility, control, and ownership for each of the processes depicted in Financial Dynamics. Fig. 5.3 describes the way in which traditional financial accounting perceives ownership and value flow. Behind each value flow is a set of processes designed and implemented by management. It is these processes that are the ultimate focus of strategy in response to competitive and technological constraints in the firm and its environment.

Figure 5.3 Value map showing how financial accounting perceives the value chain

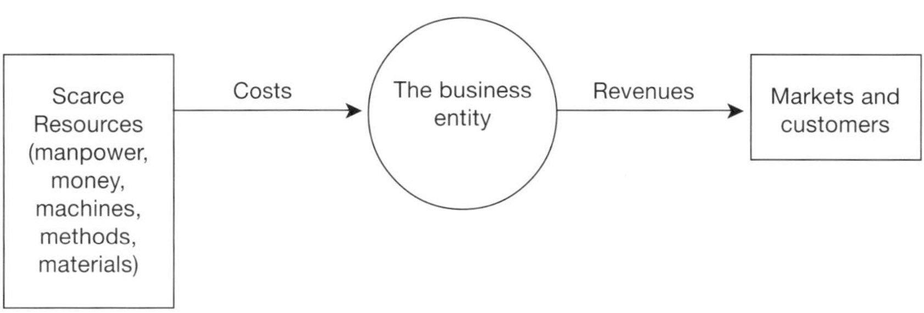

The process bubble in Financial Dynamics serves as an identifier for value flows; our tendency in normal conversation is to refer to these by their owner or manager (Fig. 5.4), since we usually tend to speak of business models in terms of the processes that comprise them – for example, assembly, programming, finance, and so forth. Of concern in verifying that a particular model of the value map is credible is assuring that we are not imposing bottlenecks (or, conversely, excessive build-up of inventory, with consequent spoilage and obsolescence costs) into our financial assessments of a process. Where the process connects a sequence of value flows – a value chain – it also displays a statistic indicating whether or not to expect idle capacity or bottlenecks in a particular process. Where the value chain spans several firms in an industry, firms at the bottleneck may be able to extract additional value, an idea that has already been used to some extent in stock-picking.

Figure 5.4 The Anatomy of a Process

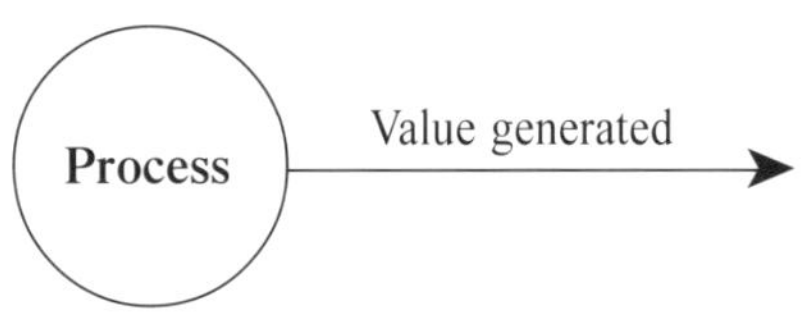

Financial Dynamics does not impose strict constraints on the quantities of transactions that may flow into and out of a particular responsibility center. The likelihood of imbalances (either excess or shortage) occurring can be tested by comparing the means of the input and output transaction quantities.[9]

EXTERNAL MARKET ("BOX")

It is conceivably possible to extend the value chain backward and forward to ridiculous lengths, though with rapidly decreasing returns and an annoying increase in complexity of the analysis. It is more reasonable to include in the value map only those processes that we are either directly interested in valuing, or that influence the processes in which we are directly interested. This is the reason for the external market box – to set boundaries on the value map. External markets also set constraints on transaction quantities and costs, which influence the value flows out of subsequent processes (for example, supply or consumer markets). A value flow extending out from a box must not exceed the constraints entered in the box – e.g., the value of programming effort (wage times man-hours) must not exceed the wages or man-hours that can be provided by the market.

The analogy to geographic mapping is clear. A regional map will provide detailed information about the region; probably a few snippets of information about bordering territories and directional information about the rest of the world. Since the user of the map specifically wants to understand the geography of the region, the "boundaries" help by culling irrelevant information from the user's field of view.

THE OWNERSHIP-ENTITY PROBLEM AND THE OFF- BALANCE ENTITY

Though Y2K failed to bring global doom on us, it *did* introduce into our vernacular the 'pregnant chad,' and the 'special-purpose entity'. The first was made infamous in the 2000 Florida presidential balloting, while the latter grew to infamy following the collapse of Texas based Enron. Special-purpose entities (SPEs) are a sort of non-consolidated, off-balance-sheet vehicle that have some legitimate uses, such as, for example, the financing of a research and development partnership with another company. They can, however, also be used to shove liabilities and risks into corners where, with luck, nobody will see them. Financial accounting's unique perspective on "ownership" makes such fudging possible.

Another example of where "ownership" concepts break down are with so-called commitments, a variety of contingent liability. A company commits itself to a future contingent payment but does not account for the liability. Telecoms-hardware manufacturers, for example, often guarantee bank loans to important customers in return for buying their products. That is fine so long as the business is healthy, but if a company enters into such a transaction purely to lubricate its own cash flow, the commitment can become a risk. Thinking in terms of "assets" rather than "value flows" allows such fudges to be used with impunity.

Yet another example where "ownership" concepts break down is in operating leases. These give a company an obligation to rent an asset over a substantial period of time, while keeping the asset off the balance sheet. Airlines use them to clear their accounts of large numbers of aircraft. The practice helps to avoid tax, but it also results in a drastic understatement of the airlines' debt. In 1997, one-third of the aircraft of the five biggest American airlines were treated as operating leases, rather than as assets.[10]

Needless to say, this is all done in accordance with accounting rules.

Value flows are less subject to these limitations, because legal issues matter less than actual flow – including arm's-length transactions.

Unlike off-balance-sheet financing methods, most of which date from the last 15 years or so, the trick of recognizing revenue too early (or of booking sales that never materialize) is an old one. Global Crossing, a bankrupt telecoms company which came under investigation by the Federal Bureau of Investigation (FBI) and the Securities and Exchange Commission, has been accused of swapping fiber-optic capacity with a competitor as a way to manufacture revenue. And according to the SEC, Xerox, a photocopier company, wrongly accelerated the recognition of equipment revenue by over US$3 billion.

Standard-setters admit that no country has adequate rules on the recognition of revenues. A solution in the meantime may be to look at cash, which is far harder to disguise or invent. Comroad duped its auditor about its revenues, but it could not conceal the fact that its cash flow was negative. This is perhaps one of the greatest reasons for cash flow being chosen as a desirable alternative to revenues in valuation. But a better solution would be to clean up concepts of what is "owned" and when that "ownership" initiates.

REPRESENTING INVESTMENTS AND ASSETS IN FINANCIAL DYNAMICS

Traditional financial investments are often couched in terms of non-perishable assets – for example, gold, securities, and so forth – which may change in value, but have a low probability of vanishing. In this context, the investment is expenditure in exchange for ownership in the non-perishable asset. But this definition of investment is much less applicable in the knowledge economy. Witness the thousands of dot-coms that went to the market looking for "investments," then promptly burnt through the funds and vanished, never to be seen again.

In GAAP, assets are unexpired expenses. They expire either through *direct consumption*, through *depreciation* (reflecting wear and maintenance), or through *obsolescence*. In the knowledge economy, obsolescence has grown in influence, while depreciation has become less relevant.

There is an entire spectrum of accounting related to investments in a firm, project, or technology. Investment expenditures can:

- buy assets that can be sold at some future date at a discount (airplanes are one such investment);
- buy assets that are unique to a project (that is, that are "sunk" costs and cannot be recovered);
- cover expenses of running a business (for example, salaries, advertising, and so forth); or
- be wasted or misdirected (for example, on speculative purchases unrelated to the core competencies).

Generally, in order to qualify as an investment, assets must generate value into the future. The fourth item was included because there is often a lag in recognizing investments that have gone sour. (For example, real estate "investments" will often be held at historical cost long after a property bubble has burst.)

The indeterminacy of either the salability, or even the usefulness, of some expenditure has traditionally created problems in firm valuation. Maintaining assets at historical cost most assuredly overstates them. But analysts have several other alternatives for valuing assets. They may be considered: (1) sunk costs, and thus irrelevant; (2) historical costs that might be sold at that value or one adjusted by depreciation (though depreciation calculation can be arbitrary); (3) replacement value (which overstates more even than historical cost); or (4) "fair value" (that is, current market value) as has been recommended by the U.S. SEC.[11] In fact, because of the general confusion about how to include fixed assets in firm valuations, no less an authority than Benjamin Graham[12], guru of the value investors such as Warren Buffett, recommends buying a diversified basket of stocks for the value of net current assets alone – that is, give no value to long-term assets such as property, plant and equipment – simply because these are likely to be sunk or, worse, subject to manipulation.

CASE STUDY: TO CONSOLIDATE OR NOT TO CONSOLIDATE ENRON'S SPECIAL-PURPOSE ENTITIES

Like most people, before their problems became public I took Enron's public pronouncements – as well as their financial statements – at face value. And I, for one, truly wanted to believe that Jeff Skilling had discovered a generic system for electronic markets which was applicable across products. These sorts of business-to-business (B2B) exchanges were a major topic in my earlier book, *Global Electronic Commerce*.

It is clear that at least some of Enron's seamier practices have survived the company's demise. An example is the so-called dark swap – the round-trip commodity trading of unused broadband (known in the telecoms colloquially as "dark fiber"). The technique is a variation on the "wash sale" in a stock market, creating the appearance of market activity and ramping up price with fabricated demand. It is strictly illegal in stock markets, which invariably install sophisticated monitoring to sniff out related party sales. The reviled WorldCom was able to cover up its operational failings with this technique, staving off default until that default was of truly gargantuan proportions.

I don't actually believe that Jeff Skilling started out with anything less than good (if not over-ambitious) intentions. Certainly, the subsequent cover-up is reprehensible, but if Skilling had wanted to bilk the public, he is clever enough to have avoided such a messy and complex denouement.

Instead, what is more probable is that Jeff Skilling (perhaps over-enthusiastically and uncritically) championed Enron's vision of electronic B2B markets. He did so in the context of intense competition and politics within Enron. Skilling had initially argued that electronic markets only work with fungible commodities (that is, each unit is interchangeable with another), until some of Enron's information technology staff came up with a "principal-intermediated" model of trading (similar to the NASDAQ stock trading system) which allowed non-fungibles such as electricity and bandwidth to be traded. In this model, Enron was forced to take large positions (perhaps 10–50% of the market) in contracts for perishable commodities (for example, a contract for electricity on Sunday has effectively perished by Monday). This "principal-intermediated" model of trading system was able to make markets in electricity and bandwidth; a truly "open platform" system, which merely searched through orders and matched up buyers and sellers, could not make markets in such contracts because they are too illiquid.

The problem is that by taking positions in these contracts, Enron also shouldered the risk of the contracts expiring without ever being sold. Even a small number of expirations could wipe out the profits from other trading. Such "principal-intermediated" markets require specialization and an intimate understanding of demand for the commodity. For example, NASDAQ dealers and NYSE specialists typically specialize in only a few companies they know well. Enron probably had this intimate knowledge of demand in oil and gas contracts, but it is doubtful they had it across the 1,000 or so markets in which they dabbled.

Add to this a lot of young, ambitious, and adrenaline-charged traders, and a culture that, according to Enron's 2000 annual report, says it "is laser-focused on earnings per share," and the risk simply grew out of control. Enron's trading losses were not a matter of "if" but of "when." I am sure that Skilling realized he was in over his head when he stepped down in the summer of 2001.

The evidence of problems had been there for some time. EnronOnline's transaction volumes grew at 60% in 1999, and 30% in 2000 and 2001, with declining profitability of each transaction. Revenues grew by US$10 billion from 1998 to 1999, and then jumped by another US$60 billion to US$100 billion in 2000. Profits before tax, on the other hand, rose by US$1 billion in 1998, and by under US$500 million in both 1999 and 2000. Enron's return on capital was only 6.6% in 2000, less than rivals such as Williams and Dynegy.

Enron delayed its day of reckoning by offloading risk to "special-purpose entities". These were not wholly independent of Enron, but were judged sufficiently separate that their profit or loss did not have to be consolidated into the company's results. Assets – that is, contracts for perishables or portfolios of perishables – were then "sold" to these entities.

Reckoning came suddenly. In October 2001, Enron wrote off US$1 billion on water distribution, broadband trading, and other investments, and suffered a US$1.2 billion capital reduction from "hedging" (that is, a ploy to lock in profits through reducing the volatility of a portfolio by reducing the risk of loss). Much of the subsequent investigation has centered on the cover-up of Enron's problems, rather than the structural problems that led to them. But until we learn from Enron's mistakes, there can be no guarantee against a similar debacle in the near future.

The question which both Enron's audit committee (headed for over 10 years by Stanford accounting professor Robert K. Jaedicke, a highly respected academic and dean of Stanford's Graduate School of Business from 1983 to 1990) and their auditor Arthur Andersen should have asked all along was, "What exactly were the transaction flows into and out of these special-purpose entities (SPEs)?" With the luxury of hindsight, and a great deal of investigation in the wake of Enron's collapse, we might also add that they should have looked specifically at two SPEs: *Chewco Investments*, a limited partnership formed in 1997 and named for the hairy *Star Wars* character Chewbacca; and *LJM Cayman SPE*, ostensibly set up to hedge Enron's substantial investment in Internet service provider *Rhythms NetConnections*, but which ultimately played a central role in chief financial officer Andrew Fastow's numerous personal scams.[13] In both cases, had Fastow actually drawn a value-flow diagram for Dr. Jaedicke and the audit committee (why would he want to?), it might have looked something like Figure 5.5.

Figure 5.5 Enron's claimed value flows

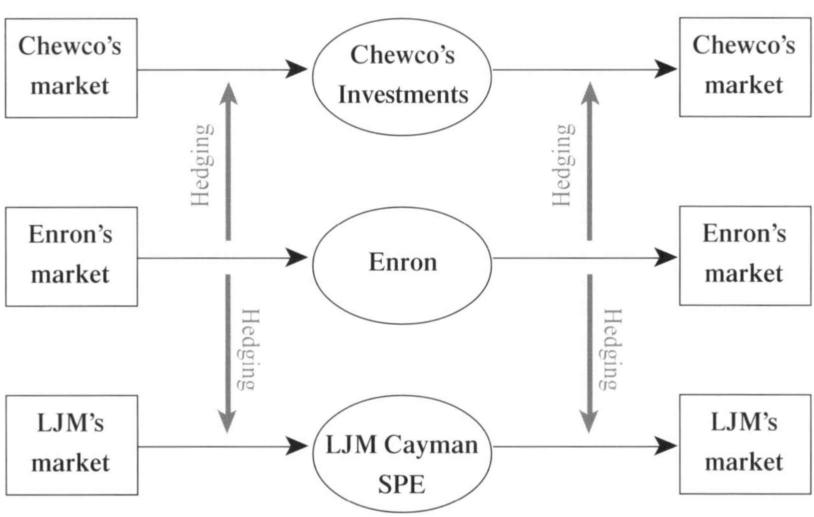

The “hedging” arrows indicate that the transactions are somehow linked with and justified by transaction streams central to Enron’s operations. Had audit committee chair Jaedicke actually been paying attention, he would have redrawn this diagram in a more realistic light (as shown in Figure 5.6).

Figure 5.6 Enron’s actual value flows

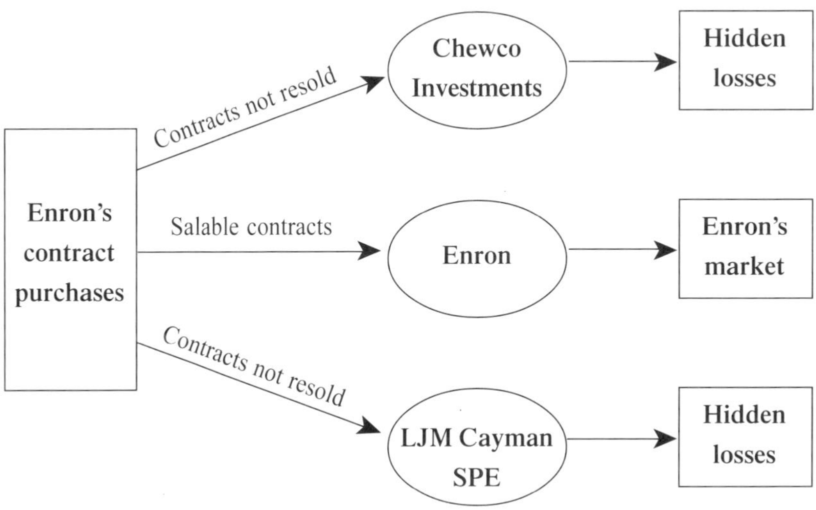

In this diagram, the appropriate question for Andersen and the audit committee to pose is a choice: “Do we consolidate the losses with Enron’s profitable contract sales; or do we turn our heads until these losses grow too big to ignore?”

Eventually, the audit committee’s failure to look at transaction flows – that is, at the dynamics of the value-creation process – was directly responsible for the greatest accounting scandal in history. Enron’s scandal ultimately wiped Andersen – once a paragon of accounting practice – from the face of the earth.

CASE STUDY: COKE, OWNERSHIP VERSUS CONSTRUCTIVE OWNERSHIP[14]

Early in this chapter, we reviewed the roots of "ownership" and how specific legal property rights define an accounting entity. Unfortunately, in a global economy, this leaves numerous loopholes – ones that are regularly used and abused – to manipulate transfer prices, barter tangible property or intangible services, and to obscure true "constructive" ownership. It is not the legal document that is important from our perspective of tracking value generation. It is the ability to influence wealth creation, and to appropriate portions of that wealth for one's own.

To illustrate the problems that arise when "constructive" ownership is ignored, let's consider the relationship of the American icon Coke to its 38%-owned affiliate, Coca-Cola Enterprises Inc. (CCE), the huge bottling company it spun off in 1986, and other members of its global bottling network. Coke doesn't consolidate CCE because it owns less than 50%. Nor, it insists, does it control CCE.

Considering transaction flows at the most cursory level, Coke obviously must control CCE. Where else, besides Coke, can they buy their flavoring syrup, and at a price set by Coke? Corporate Coca-Cola lore has it that not more than seven people in the world know the formula for the syrup at any one time, with the only written copy sealed in a safe in Atlanta. These are the crown jewels; and they are intangibles: a *recipe* and a *brand*.

Coke doesn't consolidate CCE's operations, so the capital-intensive bottler's debt stays off its books. Look at how Coke, with its lucrative and captive market for syrup, fares with respect to its debt-laden bottlers. (See Figure 5.7 comparing the stock price returns of the two companies.) In the 1990s, Coke booked hundred of millions in annual payments to CCE as assets, not expenses, engineering their profit over a longer time; while at the same time it relentlessly pushed bottlers to expand (creating more demand for syrup) while their profits collapsed.

Figure 5.7 Stock Returns for Coke and CCE

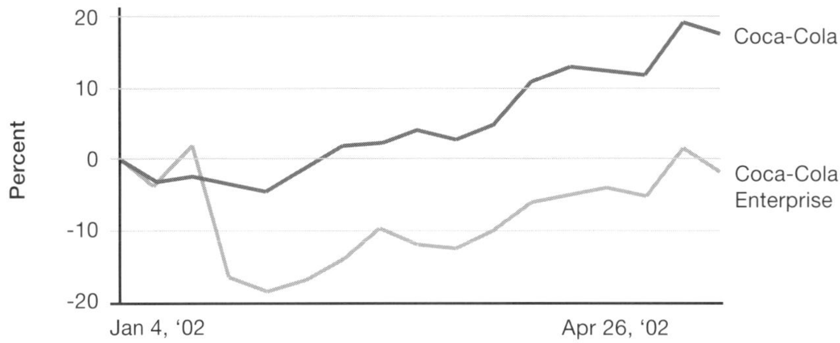

Source: Bloomberg Financial Markets

At their annual meeting, Coke chairman Douglas N. Daft boasted: "Our financial results speak for themselves." But do they? Table 5.1 shows unconsolidated and consolidated returns over the past six years.

Table 5.1: Return on Assets for Coke and CCE

	1997	1998	1999	2000	2001	2002
Coca-Cola Co	21.5%	18.3%	14.7%	17.2%	17.5%	17.8%
Bottlers	2.8	2.2	1	1	1.9	2.8
Combined	8.8	7.7	5.9	6.3	7.5	8.6

Source: Lehman Brothers Inc.

The fact is, Coke made high returns in the 1990s in part by exploiting its bottlers. But in the future, Coke will have to help them repair the damage to their balance sheets that resulted from forced over-expansion. And that will likely cut into overall earnings growth going forward. All of this recognizes that value lies in future transaction flows. And it recognizes, contrary to generally accepted accounting myth, that Coke and CCE are financially joined at the hip.

---

[1] *Ralph Waldo Emerson (1850)* Representative Men, reprint, New York: Judy Henteloff; December 1980, from the chapter "Montaigne or, The Skeptic"

[2] Quoted in Hesketh Pearson, *The Smith of Smiths* Chapter 11; reprint by New York: Scholarly Press (June 1989).

[3] Steven Pinker, *How the Mind Works* (New York: Norton, 1999), Chapter 3.

[4] W. Baumol, *The Free-Market Innovation Machine* (Princeton, N.J.: Princeton University Press, 2002).

[5] "Rule of thumb' comes from an old British common law rule which, supposedly, allowed husbands to beat their wives with a stick no thicker than their thumb, though there is considerable debate over whether or not this law really existed. Blackstone, in his influential 'Commentaries on the Laws of England' (1765), wrote that "The husband also (by the old law) might give his wife moderate correction." The 'rule of thumb' itself has been first attributed to Lord Robert Raymond, who was a judge on the King's Bench (a common law court) in England from 1724-1733.

[6] In theory, if markets were perfectly efficient, these two values would converge. But markets, like other human inventions, are less than perfectly efficient, and introduce their own idiosyncrasies into stock and bond prices.

[7] The value chain is a conceptual model of the value transformation in firms which was introduced by Michael Porter in his 1985 book. Competitive Advantage. Porter's idea derived from the economic intuition of production that every firm is a collection of activities that are performed to design, produce, market, deliver, and support its product. All these activities can be mapped using a value chain. A firm's value chain and the way that it performs value-adding activities reflect its history, its strategy, and the underlying economics in its markets.

[8] There is a rule of thumb that control processes can be expected to consume 60–90% of the cost of operating a business system, but to my knowledge, there has been no empirical evidence to support this range of figures, though executives with whom I've discussed this have commented that this agrees with their experience, and appears to be a reasonable range.

[9] Because the focus of Financial Dynamics is on accurately estimating the mean, which implies that we want our estimators to be unbiased, the variances are not specified. This creates a difficulty in comparing two means called the Behrens-Fisher problem. The Welch correction is a widely applied "solution" to the Behrens-Fisher problem resulting from the inequality of the variances. When the variances are not equal, the Welch correction is applied to modify the acceptance region of the test (see Scherrer, 1984 and Zar, 1999). The Generalized Likelihood Ratio Test statistic for this condition when *the transaction volumes* are normally distributed is the *t*-statistic. B. Scherrer, *Biostatistique* (Chicoutimi: Gaëtan Morin, 1984); Jerrold H. Zar, *Biostatistical Analysis,* 4th edn (Englewood Cliffs, N..J. : Prentice-Hall, 1999). I should also note that the Behrens-Fisher problem does not appear in Bayesian inference, which therefore can avoid the previous 'corrections' altogether.

[10] Source: Trevor Harris, at Morgan Stanley Dean Witter.

[11] Jackson Day, 28th Annual national conference on current SEC developments, 2000 (Jackson Day was deputy chief accountant for the U.S. SEC).

[12] Warren E. Buffett (Preface), in Benjamin Graham, *The Intelligent Investor: A Book of Practical Counsel*, 4th rev. edn (New York: HarperCollins, 1985).

[13] Enron's 217-page internal report on their own financial misdealings states that Enron's chief financial officer Fastow used his position in Enron to influence (or attempt to influence) Enron employees who were engaging in transactions on Enron's behalf with the LJM partnerships. The report detailed how Fastow and other executives apparently enriched themselves from the questionable partnerships they helped to create. In one case, the former Enron CFO allegedly made US$4.5 million from a US$25,000 investment – in two months. Kopper, one of Fastow's employees, is said to have earned at least US$10 million from a US$125,000 investment in one of the partnerships, the report notes.

[14] Data was reported in "How Coca-Cola Uses its 38%-owned Bottling Company, CCE, to Improve its Financials," *Business Week*, May 13, 2002 (Dean Foust and David Henry).

CHAPTER 6

Financial Dynamics

All models are wrong, but some are useful.

– G. E. P. Box,
Robustness in Statistics (1979)

EXISTING VALUATION METHODS

Analysts employ a variety of approaches to value a business or project. Some of these have been designed to handle specific jobs. For example, a price-earnings ratio allows a quick-and-dirty computation of a firm's value from annual income. In general, the character of the valuation process depends on who is doing the valuation, and what their objectives are.

To make my point, consider the contrast in objectives between the Lehman Brothers debt and equity analysts in the summer of 2000 when the firm was constructing valuations of bookseller Amazon. Lehman Brothers debt analyst, Ravi Suria, predicted that Amazon would fail based on cash flow (which directly impacts Amazon's ability to repay debt). On June 23, 2000, Suria issued the following warning: "We believe that the combination of negative cash flow, poor working capital management, and high debt load in a hyper-competitive environment will put the company under extremely high risk."

The report pushed Amazon to an 18-month low of $33 7/8. In six months, its market value dropped from US$40 billion to around US$10 billion. Suria's counterpart in equities, analyst Holly Becker, stood by her "buy" recommendation for Amazon throughout the Suria-induced price drop.[1]

Predictably, Jeff Bezos, CEO of Amazon, dismissed Suria's predictions as "hogwash," a sentiment that was reiterated by Amazon spokesman Bill Curry. "Anyone who understands the cash flow dynamics of this business knows this." Who was right? It's hard to tell, since Suria's dire forecast never came about. Amazon changed course smartly, got its costs under control, and tamed its aspirations of "growth at any cost." Yet Holly Becker's "buy" was equally flawed. Throughout 2001, Amazon traded under $10, which would have dashed hopes of any quick gain even for those getting in after Suria's warning.

The moral? Things change. And valuation methods need to account for change, by providing clear roles for assumptions and existing data, a scientific approach to inferring the future impact of strategy and the associated risk, and the ability to seamlessly alter a prediction whenever new information becomes available.

During the boom years of the 1990s (and one might note in tandem with the rise of extreme sports) valuation took on the character of a spectator game, complete with side-bets. But whereas extreme sports have thrived in the new millennium, analysts find themselves defensively trying to account for their activities and value assessments during the boom years. The mightiest have been tarred and run out of town in disgrace. At least there is now the wide consensus that valuation must drag itself out of the arena of dueling egos, and find a more objective, complete, and scientific foundation.

Table 6.1 details the main valuation approaches in use at the end of the 20th century; it identifies where they are used, and what are their strengths and weaknesses. I have thrown in Financial Dynamics as well for comparison, showing how it can improve on our current standards for objectivity, completeness, and effective use of scientific tools.

Table 6.1 Valuation approaches: Origins, benefits, and defects

Valuation Method	Origin	Claimed Benefits/ objectives	Defects
Financial accounting (accruial basis evolved in 20th century)	12th century formalized in 15th century	Objectivity	Incomplete model of the modern firm, perhaps less than 20% of value generating activity today is tracked by financial accounting.
Cost accounting	Early 20th century, part of Taylor's 'scientific management'	Risk, budgets and targets are included in the model	Incomplete model of the modern firm, specifically directed toward industrial production
Managerial accounting	1950's at Procter & Gamble	Risk, budgets and targets are included in the model	Incomplete model of the modern firm, specifically directed toward discretionary spending
Discounted Cash Flow (WACC, etc)	1938 in The Theory of Investment Value by John Burr Williams	Time is explicitly included in the model	The wide latitude for individual interpretation of the method makes valuations incomparable and inconsistent across individuals, ad hoc treatment of discount rates and inflation create similar inconsistencies across time
Price-level Adjusted DCF (used in Boston Consulting Group, Collins Stewart and HOLT Value Associates Cash flow return on investment (CFROI) & Total Shareholder Return (TSR)	1940's during depression era	Time and relative price levels are explicitly included in the model	Same problems as DCF, with the added uncertainty of finding deflators for price level
Residual Income (used in Stern Stewart Economic Value Added (EVA) and Market Value Added (MVA); L.E.K.share-holder value added (SVA); Marakon Associates economic profits (EP), KPMG economic value management (EVM), economics profits (EP), market value added (MVA, etc)	1950's at Procter & Gamble	Time and historical costs for future benefits (i.e., investments) are explicitly included in the model	Same problems as DCF, but less extreme. Because there is no need to compute cash flow forecasts from earnings, the forecasts and resulting values tend to be more precise
Financial Dynamics	Early 21st century	See text	See text

In brief, the techniques available for valuation consist, on one hand, of accounting and its cost and managerial extensions; and on the other of discounted value flow variations. The strengths and weaknesses of these approaches were explored in depth in my prior book, *Valuing Technology,* Chapters 7 and 8.

Accounting is a process of classification of economic events, and the summation (with double-entry for error control) to compute valuation statistics. The quality of accounting valuations is only as good as the classification system. (Accounting's taxonomy is called its "chart of accounts.") This classification system determines what data (that is, transactions or economic events) are gathered by the accounting system; valuation is set at the price of the last arm's-length transaction. This classification system was more or less formalized at a time when world economies ran on agricultural products, and has become hopelessly outdated in the knowledge economy.

Discounted value flow (DCF) methods look to the future rather than the past, then try to figure out what those future value flows are worth today. For unadorned DCF, the future value flows are some forecast of net cash flows into or out of the company. With price-level adjustments, these are adjusted by some "price index," a process which itself can be controversial and subjective. Residual income approaches are hybrids of accounting and DCF that eliminate the uncertainty in future forecasting. None of these models explicitly prescribes how future values should be forecasted, which allows considerable latitude in performing the actual calculations, a point which I argued in detail in my book, *Valuing Technology*.

Generally, the best way to derive better valuations is to get better data. The old saws that "you can't make a silk purse out of a sow's ear" and "garbage in, garbage out" have a converse: "the first step in making a silk purse is to buy some silk." In this book, I take accountants to task repeatedly, not because they are the source of valuation problems, rather because they are the first step in good valuation. Along with securities analysts who keep an overly cozy relationship with their corporate subjects, accountants are the only third parties with direct access to the data required for accurate valuation.

Table 6.2 describes data sources, methods, and output (or absence of them) for the valuation approaches in Table 6.1.

Table 6.2 Valuation approaches: Data requirements and modeling conventions

Valuation Method	Source Data	Prediction Method	Time-Range Domain	Aggregation	Error Control	Risk Measures
Financial accounting (accruial basis evolved in 20th century)	Historical 'Arm's Length' (i.e., with external markets) Transaction of Tangible Goods	None (predictions are not considered 'objective')	Historical transaction	Summation	Double-entry	None
Cost accounting	Historical transactions, internal and external in tangible goods	Target prices/ costs set by strategy	Historical transaction; next-year target for std costs.	Summation	Double-entry	Variance serve as basis for management by exception
Managerial accounting	Historical transactions, internal and external in tangible goods	Target prices/ costs set by strategy	Historical transaction; 1-, 5-, 10-year plan	Summation	Double-entry	Variance serve as basis for management by exception
Discounted Cash Flow (WACC, etc)	Forecast Cash flows based on financial accounting statements	ad hoc with wide latitude for individual interpretation	2-stage: (1) growth 5-15 years; (2) plateau	Summation with discounting over time and ad hoc determination of discount rate	Double-entry (assuming financial statements are the data source)	None. Risk may be incorporated in the calculation for capital cost and for future value
Price-level Adjusted DCF	Forecast cash flows adjusted by inflation and holding gains/ losses – i.e., market values	ad hoc with wide latitude for individual interpretation	2-stage: (1) growth 5-15 years; (2) plateau	Summation with ad hoc discounting and price-level adjustments over time	Double-entry	None. Risk may be incorporated in the calculation for capital cost and for future value
Residual Income	Forecast financial accounting earnings and historical costs for asset investments	ad hoc with wide latitude for individual interpretation	2-stage: (1) growth 5-15 years; (2) plateau	Summation	Double-entry	None, though risk may be incorporated in the calculations for capital cost and for future value; note also that Marakon and L.E.K. add their own risk assessments
Financial Dynamics	Forecast financial accounting earnings plus forecast earnings from ideas and strategy; assets are considered sunk cost and ignored	ad hoc with wide latitude for individual interpretation econometric process modeling fitted to historical data	Unconstrained	Statistical approaches to estimate strategic parameters	Extensive controls at point of data entry (double entry is abandoned)	Statistical distribution of the forecast estimate; can be used to calculate estimator dispersion, Bayer risk, Markowitz risk premium, average or relative risk aversion

Better data does not mean *more* data, a fact that has more than once confused the finance community. In a quest for greater accuracy and authenticity, DCF spreadsheets sporting detailed forecasts have been embellished with things such as depreciation, sales returns, inventories, and whatnot for 10 to 15 years into the future. More line-items means more precision, right? Or is this just the 20th-century update of shamanic incantation – ritual recitation of verbal charms and spells to produce a convincing illusion?

In Table 6.3, I have listed some important valuation alternatives that I have not included in the detailed lists, and an explanation of why they are omitted. There may indeed be reasons to reconsider; but for the most part, these have special uses, and as a result are not generally applicable to the valuation of business.

Table 6.3 Special-purpose valuation methods not included in comparison

Not included, not directly comparable	Why?
Real options, decision trees and Monte Carlo approaches	Objective: Pricing uncertain alternatives; better for simple projects with a few alternatives
Bond ratings; Debt Oriented Valuations (Moody's, S&P, D&B)	Objective: to determine ability to repay debt, often dimensionless, e.g., AAA
Value -at-risk	Objective: Potential loss if adverse event occurs.
Return on investment	Dimensionless 'rate' value does not measure absolute value of firm; value of 'investment' is nearly impossible to determine in knowledge businesses such as software where there is no clear delineation between 'asset' and 'expense'
Price-earnings and other quick and dirty approaches	Incomplete theory – dependent on faith in 'market efficiency'.
Actual vs cash flow analysis	Really the same, but accrual anticipates cash flows prior to the actual exchange of 'tangible' property
Market based methods (CAPM, Arbitrage-pricing theory, etc)	Notoriously imprecise, and dependent on assumptions that are suspect in the knowledge economy

SOAP OPERAS AND MANAGEMENT ACCOUNTING

Harley Procter and James Gamble began business in Cincinnati in the 1870s selling candles and soap from a wheelbarrow. James Gamble, an inveterate chemist and tinker, perfected his formula for "P&G White Soap" in 1878. Unsatisfied with this appellation, Harley Procter decided to give the soap a name that people could remember – he named the soap "Ivory" after the Psalm 45 verse: "All thy garments smell of myrrh and aloes and cassia, out of the ivory palaces whereby they have made me glad."

Gamble's floating soap was invented by accident. Without thinking, one of the P&G factory workmen left to go to lunch while the machinery was still running, letting air work its way into the mixture. Procter & Gamble began to receive letters from buyers of this accidental soap. They wanted more of the soap that floats! The product was popular because Cincinnatians often bathed in the Ohio River (this *was* the 1880s) and the floating soap would never get lost.

Procter wanted to produce pure soap. Unfortunately, a chemical analysis of Ivory soap found that 0.56% of the ingredients did not fall in the "pure soap" category. In 1882, an Ivory advertisement appeared that stated the slogan "99-44/100% Pure." This calculation was created by Procter in which 56 was subtracted from 100 to account for a result of 44. The 0.56% consisted of the following impurities: uncombined alkali, 0.11%; carbonates, 0.28%; and mineral matter, 0.17%. "99-44/100% Pure" became one of those advertising memes – like "LSMFT" and "Where's the beef!" – that took on a life of its own.

Well, none of this naturally had anything to do with the soap's main function: to get you clean. Instead, it was one of the first deliberate exercises in branding. P&G became one of the first corporations whose main investment was in an intangible asset – their "brand equity." P&G's branding exercises were exceptionally successful over the next 50 years, as it grew to be one of the largest, most successful, and most powerful firms in the world.

The corporate culture of America's most respected firm typified the 1950s' "Corporate Man" mentality. When Harvard Business School restructured itself in the 1950s, it based its curriculum and departments on the functional "stovepipes" at P&G – accounting, marketing, finance, strategy, and human resources – setting a standard for American business schools that has endured to this day. P&G is still the largest U.S. maker of household products, with nearly 300 brands in five main categories: (1) baby, feminine, and family care; (2) fabric and home care; (3) beauty care; (4) heath care; and (5) food and beverage. It also makes pet food and water filters. Twelve of P&G's brands are billion-dollar sellers (Always/Whisper, Ariel, Bounty, Charmin, Crest, Downy/Lenor, Folgers, Iams, Pampers, Pantene, Pringles, and Tide).

The unique character of a company whose main business was in intangible "brands" required an innovative approach to accounting and management. Financial accounting was inadequate, and cost accounting was only suitable for tracking production processes. P&G's real wealth was in its brands. Advertising, brand management, resellers, retailers, and customer relationships were paramount, and thus garnered the lion's share of investment and expenditure at P&G. For these purposes, P&G invented managerial accounting, with its focus on discretionary budgets, breakeven analysis, and consumer markets. How else could they have justified the production of soap operas – a genre invented by P&G – such as *The Guiding Light* and *As the World Turns*? It is also interesting to note that residual income approaches for valuing public relations projects and product lines were first used regularly at P&G in the 1950s.

A half-century has passed since P&G rethought accounting for a business based mainly on "ideas" and "intellectual property." Much has changed – ideas can now take life, in software and robotics. Tangible products such as cars and disk drives have become so refined that much of the work on them can only be done by machines – there is no labor–capital tradeoff. Strategy is ready for new accounting and valuation tools. In the next section, I lay out the components and methods through which Financial Dynamics achieves its objectives.

THE FINANCIAL DYNAMICS METHOD

I have provided a step-by-step summary of Financial Dynamics valuation in Appendices A and B. All of the work in Financial Dynamics takes place on the arrows in the value map – the value flows that percolate through the value chain. It is here that data are collected – some from financial statements, and some from auxiliary sources. Financial Dynamics' forecasting "black box" adopts a formal econometric approach to predicting the future value flows. This has the advantage over existing DCF and residual income approaches of transparency, replicability, and statistical validity. Because of this, Financial Dynamics' forecasts are more informative, less subject to analysts' biases, and less volatile than the *ad hoc* forecasts used in the existing valuation approaches described in tables 6.1 and 6.2.

Financial Dynamics partitions the source of value into three categories: those attributable (1) to *strategy*, (2) *technology choice*, and (3) *environment* (Fig. 6.1). *Strategy* is presumed to be actively managed by the firm. *Technology choices* are made by the firm, but almost all of the development in the technology takes place outside the firm. Because technology knowledge is a public good, the firm can borrow whatever it needs from the public 'knowledge base' at no charge. But the only part of the public knowledge base that is useful is constrained by the firm's original technology choices. Management can influence the *environment* only through strategy and technology choice. The *residual environmental factors*, for good or bad, are an endowment that the firm cannot avoid.

Figure 6.1 The Financial Dynamics Value Space

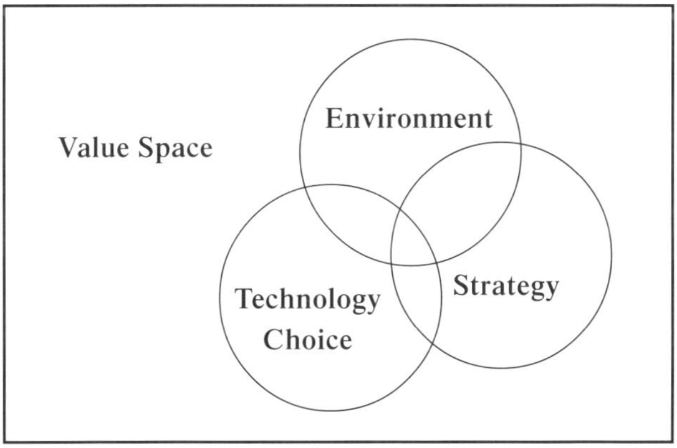

Through particular strategy and technology choices within their competitive environment, management makes choices about what proportion of value is generated now versus what part they will defer to the future. The value – call it v – generated by operations in any particular period is the result of controllable strategy and technology choices, and uncontrollable environmental and risk components

We compute our forecasts in Financial Dynamics from an estimation period of *historical* values generated by the company, and for *historical* policy variable values for the strategies and technologies that management *plans to use in the future*. This allows us to investigate theoretical future situations: '*what if' these* strategy and technology choices were made in the environment we have historically operated in. Assume that the things we expect to change in the future are controlled by technology and strategy choice; the things we expect to stay the same in the future are the residual environmental factors (a reasonable assumption if there is no additional information available on the environment). Taking one year as our standard period, let that value be the portion of value generation that is delayed for one year, rather than being consumed in the current year. The proportion of next year's value that is also realized this year and is attributable to a particular set of strategy and technology choices is some value – call it ρ; the proportion of next year's value that is unrealized this year (i.e., deferred by choice by one year) and attributable to a particular set of strategy and technology choices is $1\text{-}\rho$. The value $1\text{-}\rho$ is management's 'own discount factor' on an annual basis; it is the implied rate at which value next year is traded for value today.

Appendix B provides a method to estimate $1\text{-}\rho$ that dispenses with many of the problems of cost of capital numbers used in residual income and DCF approaches. Residual income requires that future value flows be discounted with an *opportunity* cost of equity capital. Typically a *historical* cost of equity capital is computed using the capital asset pricing model (CAPM). CAPM itself has historically predicted only about a third of the variation in the stock market, and recently has yielded even worse predictions, leading many of its proponents to abandon it. Weighted average cost of capital is computed from a proportion of equity capital cost and tax adjusted debt capital cost (both historical). It requires assumptions about capital structure, effective tax rates, and debt markets well into the future that are in practice available.

Through its strategy and technology choices, management has taken a particular value $\boldsymbol{v} \times (1\text{-}\rho)$ and deferred it one year. Managements *'own interest rate' for a given strategy-technology choice* (or equivalently management's '*own discount factor*') may be computed on the basis of a particular value when it is available at different dates.[1] In the context of management's particular strategy and technology choices, then *own interest rate* $\frac{\nu \times (1-\rho)}{\nu} = 1-\rho$. Management's particular strategy and technology choices reveal that their perceived opportunity cost of capital is $1-\rho$. This makes sense where we see 'capital' as an ongoing set of infrastructure negative value flows, rather than the traditional 'stored wealth' capital that we are used to seeing on the balance sheet.

Compare this to the discount rate used in traditional residual income approaches such as economic value added (EVA). Here the *cost of equity capital* is determined through the capital asset pricing model (CAPM) and makes a fundamental assumption that the 'opportunity' given up is some investment in a portfolio of marketed securities. But management would be unlikely to buy a portfolio of shares as an alternative to the particular investment of capital associated with the chosen strategy and technology; instead it would either (1) invest in other projects, (2) buy back its own stock, or (3) issue a cash dividend to its shareholders (the latter two merely give money back to shareholders). Thus the real 'opportunity' is to invest in other projects, and this opportunity cost is best reflected in the 'own discount rate' which reflects management's own perceptions of what their opportunities would be in the context of a specific set of strategy and technology choices.

Once the future stream of value flows has been forecast, then a variety of techniques can be used to aggregate these for decision-making. In addition, for scenario analyses that require knowing not just the magnitude of value, but when it will be generated, the disaggregated forecasts will be optimally informative. Aggregation of the forecasted value streams is covered in Chapter 9.

Financial Dynamics separates the transaction volume and value flows, and forecasts them separately. This is by design, because any given strategy will impact volume and value differently. For example, a strategy of "commoditizing" a product strategically goes after high-volume and low-transaction-value production approaches; the converse targets an exclusive, low volume, "luxury" market with a high markup.

CHOICE OF THE "VALUE" METRIC

Accounting provides little information on *marginal costs and revenues* in the financial statements. Firms do not report marginal costs and revenues, and most could not if they wanted to. So, *average costs* are relied upon as a surrogate, for example, in setting sales price, in deciding which products to product, in capacity planning, and so forth, even though marginal costs are often more useful for decision making. Consider what is included in the average cost of a typical good. There is the cost of material and labor required to produce a single unit, which may be nil in intellectual property such as software. There is allocation of overhead, which will likely include fixed assets, which are depreciable, and labor, which is not. Depreciation is simply a way of allocating multi-year costs to units, though accountants usually state this differently, as allocation over useful life of the asset – useful life meaning the duration over which the asset is punching out these units. Labor that is allocated is often associated with so-called "support" tasks, which are designed to coordinate, control, or otherwise make processes operate more efficiently. Marginal costs would normally exclude these allocated costs. If these are already presumed to be "sunk" costs – that is, spent and gone, whether we use them or not – then average costs wind up providing a better approximation of marginal costs.

Cash flow has traditionally been preferred for valuation because it is considered less exposed to accounting "manipulations" – a term often used in the finance community to describe allocations of historical costs such as depreciation and overhead. But accounting adopted accruals and allocations in the early part of the 20th century amid much debate, to combat the misstatement of earnings in cash-basis accounting.

My own take on the debate is that the problem is not whether cash versus accrual is better for computing value, but whether historical costs should be allowed to forever haunt future financial statements. Unless accountants can compute their accruals and allocations accurately – and the evidence indicates that often they do not – then accrual accounting will be seen as "manipulative" by people who actually have to find a precise value for the firm and its operations.

We owe this unhappy state of affairs to two causes. First, accounting – in reverence to crop planting and reaping cycles – divides costs into those that are used up in less than a year (i.e., expenses) and those that are not (i.e., assets). Agrarian tradition, blindly pursued into the 20th-century industrial and knowledge economies, has harvested a profusion of arbitrary time allocations – that is, depreciation expenses.

Second, because accounting statements present account values as precise statistics, rather than as uncertain random variables, they disappoint when it is discovered that they lack accuracy. The degree of inaccuracy, which accounting measures crudely as the auditor's "materiality," has seldom been revealed by accounting firms, except under court order. This again contributes to the image of "manipulative" accounting by people who expected precision, because account values are stated as if there is no uncertainty.

Financial Dynamics offers a solution to both of these problems. First, rather than recording costs as assets which will depreciate, all costs are considered "sunk," thus obviating the need to depreciate them. Precedents exist for this treatment of expenditures in certain schools of valuation, in particular by Benjamin Graham who advises Warren Buffett. If, indeed, some of these "sunk" costs end up having a market value, then the Financial Dynamics approach will tend to be conservative. If investments are a large component of value, and we are assured of recovery of some of the investment should management divest, then the "salvage" value of the investment could be added to the present value of the value flows to more accurately reflect wealth. In most cases, future value flows should be large enough that this "salvage" value will remain less than one standard deviation of the estimated present value of the project or firm. In this case, it is reasonable to simplify life, and just treat the whole investment as a "sunk cost."

UNCERTAINTY AND RISK IN VALUE FLOWS

The most significant advantage Financial Dynamics offers over other approaches to valuation is in providing formal methods for assessing uncertainty and risk. [2] Traditional discounted value flow or traditional accounting methods provide only a single point estimate for value, and are mute regarding risk and error in that estimate.

A dispersion measure such as variance is, by itself, incomplete. The individual making the decision based on an estimate has to apply his decision's so-called loss-function – that is, an amount of money or utility that he may lose if the estimate is "wrong" by a certain amount. The term "loss-function" is that generally used in statistics, but in fact, the function can, and typically will, represent both losses and gains, value taken away and value added. The distribution along with the loss can be combined to compute an expected loss, or other meaningful measure of risk.[2]

Many decisions have asymmetric and non-linear loss functions. For example, consider the investment in an option to sell stock at a given price – a so-called put option. If the estimate of the stock price that brought about this investment (when the exercise price is chosen) was too high (that is, the current price is lower than the option's exercise price), then the option has a fixed value of zero; in contrast, larger errors on the low side at the time of investment have an accelerator effect, pushing the put "into the money" because stock price has risen, resulting in ever-greater gains for the option holder.

IMPLEMENTATION RISKS

Columnist Andrew Allentuck, in an excellent review of my previous book *Valuing Technology* for *The Globe and Mail*, observes that some of my suggestions for updating accounting are destined to meet resistance. He asserts that some of the suggestions "would drive securities regulators and tax collectors mad" and their complexity "would mean that financial statements could only be read by math mavens." I have no doubt that there is a substantial element of truth to Mr. Allentuck's observations. Nonetheless, I am also reassured that technology itself will ultimately come to my rescue, because computers have the potential to automate much of the complexity that is inherent in a more complete accounting. Sure, we might lose some transparency arising from our existing summations and orthodox accounting mechanics. But such a loss is not substantially different from the "transparency" we have lost in the use of Web-based searches, word processing, pocket calculators, and so forth – all of which have moved us from "transparent" paper processes to complex automated substitutes. Once the transition is made, we will most likely be surprised that we could ever value firms in this "old-fashioned," labor-intensive way. Sure, we might have to rely on computing machines, but the ubiquity of these machines is inevitable, their use becomes simpler over time, and the effect is to free us to tackle the creative problem-solving to which we are most suited.

Appendix A: The Financial Dynamics Methodology

Financial Dynamics is implemented in a sequence of four groups of activities, as depicted in Fig. 6.2

Figure 6.2 Steps in Financial Dynamics valuation

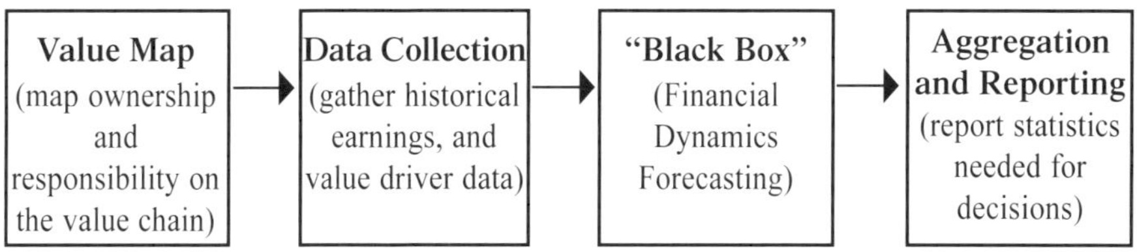

In Financial Dynamics, the degree and character of uncertainty is inferred from statistical analysis of prior value flows, and how much of the variance in these flows is attributable to managerial strategy (as formally depicted in the strategy models discussed in Chapters 7 and 8). This is done through an econometric model that is at the center of Financial Dynamics' forecasting "black box." Figure 6.3 shows how this works.

Figure 6.3 Financial Dynamics depicted as a "black box" which converts strategy and history into value flow forecasts and present values

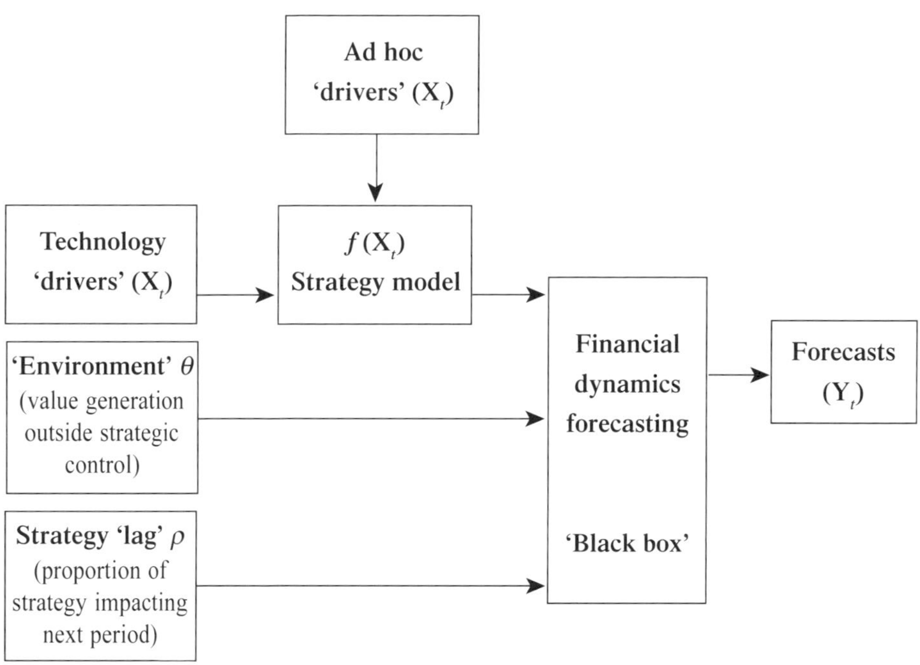

I have provided the detailed mathematics inside the "black box" in Appendix B of this chapter. At this point, I would like to focus just on the important components of the "black box." Five sets of data must be collected for the forecasts of each of value and volume on a particular value flow "arrow" and input to the "black box."

1. **Historical numbers for value and volume on a particular value flow "arrow."**
2. **Historical values of strategy drivers** for all of the strategies enabled on a particular value flow "arrow." Note that there are two types of drivers: (1) technology choice drivers that are the same for a given technology across the economy; and (2) *ad hoc* strategy drivers that are unique to the operations of the company.
3. **Planned values of strategy drivers** for all of the strategies enabled on a particular value flow "arrow."
4. **The *business environment value contribution*** (θ in the notation in Appendix B) This is the term I use to describe the value that has historically been generated by forces outside of management's strategic control. It could represent lucky choices in product or investment markets, economy-wide economic effects, general uncertainty, and it could be the product of good fortune, misfortune or difficult times.
5. **The *own discount rate*** ($1 - \rho$ in the notation in Appendix B) is the value trade-off over time implied by management's stated strategy and technology choice.[3] The calculation of $1 - \rho$ from management's own decision making process provides a simple, accurate and robust alternative to one of the great conundrums of traditional present value approaches to valuation – DCF, Residual Income, APV, and so forth. In these approaches, the discount factor is derived from capital structure (the way that the firm finances itself through the sale of claims to the firm). DCF computes a cost of capital weighted by the amount of debt and equity capital, followed by a deduction of debt claims prior to the computation of share price (assuming the question asked is 'what is a share worth' rather than 'will the firm repay its debt'). This is a problem. Since the 1970s when DCF became a standard for valuation, financing has grown steadily more creative. Because of this, the classical DCF approach distorts the valuations of firms with complex

ownership structures, especially where inflated shares prices have been used to buy into firms – de rigueur for knowledge economy firms. The approach requires managers to presume that they know the true value of those other claims. In practice, they do not know the value of all claims against the firm. To find this value of claims entails an additional valuation methodology, and cost of capital computed in this way is likely to be volatile, since capital from many sources is usually pooled because any one of the participants in insufficiently wealthy, or is unwilling to bear the risk of a potential default, or cancellation of the project. Residual income presents even worse problems because there is no widely accepted formula for computing opportunity cost of equity capital – usually the discredited capital asset pricing model is used by default, but yields unconvincing opportunity costs for equity capital. The Financial Dynamics *own discount rate* $(1 - \rho)$ recognizes that the discount of future value flows should use an opportunity cost that reflects the best assessment of alternative opportunities. Since management is presumably in the best position to make the judgment about feasible alternatives, the Financial Dynamics *own discount rate* $(1 - \rho)$ provides a very compelling opportunity cost of capital as assessed by the management of the firm.

The output of the black box is a set of forecasted amounts of value or volume on a particular value flow "arrow." (In the statistics I will call this Y_t, and for the forecasted values $t = T+1, ..., T+m$, where T is the current date. The forecasted value Y_t depends on historical and planned values of the strategy drivers as well as "environment" and "strategy lag". These inputs and outputs can be grouped into historical components necessary for calibrating the estimation model; and forecast components, necessary for generating the predictions as shown in Figure 6.4.

Figure 6.4 Estimation and prediction components of Financial Dynamics

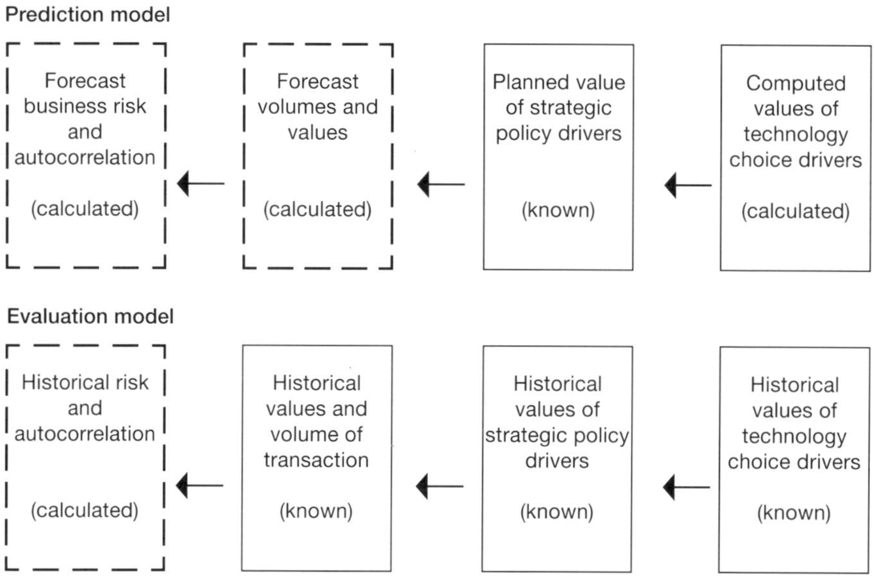

Chapters 7 and 8, respectively, will define precisely what is meant by "technology choice drivers" and "strategic policy drivers." These are critical to the success of the Financial Dynamics approach, as they quantify critical factors required for valuation, but which are normally perceived as qualitative, rather than quantitative, data. Quantification of risk requires that we quantify the major sources of risk in models that describe the impact of management's stated strategy, and the technology choices in which the firm invests.

Compare this approach to discounted cash flow. DCF aggregates historical information in the same fashion as financial accounting – thus leading to suspicions that a typical DCF analysis is today capturing only around 20% of the value generation of the firm. There is no explicit measure of forecast error or risk for the projected future cash flows, nor any specified approach to linking forecast values with historical values. Strategy and technology are only incorporated into DCF via *ad hoc* extrapolations of accounting statement line items – often only the future sales revenue is adjusted based on guesses of the impact of a given strategy or technology many years from now. Analysts using DCF as an aggregative method may in individual cases be more meticulous in forecasting specific lines, but there is no standard of performance and thus no consistency across valuations. For this reason, it is common to see analysts disagree dramatically on valuations when given the same data; and also to find analysts revising valuations significantly even though they have uncovered no new evidence. (In *Valuing Technology* I provide an example of an analysis of Amazon, which dropped to 20% of its value after only six months, and with no new data; the dot-com era was awash with such revisions.) Similar issues plague residual income approaches.

The basic inferential framework of Financial Dynamics is provided through a non-linear least squares regression of transactions with respect to a set of strategy and technology "value drivers." The intent is to capture all of the salient firm-specific factors influencing fundamental value in knowledge-intensive firms. Financial Dynamics is kept simple, in order to parsimoniously use the small datasets that are typically available. This is an effective way of reducing uncertainty in measurement (which can result in volatility, since small changes in data can produce large changes in parameters).

Financial Dynamics Step by Step

Step 1: The first step in the Financial Dynamics analysis is to draw the value map (Fig. 6.5) – the schematic of all the value flows in the system (arrows), the "owners" of those value flows (bubbles), and the "environment" boundaries of the value-generation process (boxes).

Figure 6.5 The Value Map

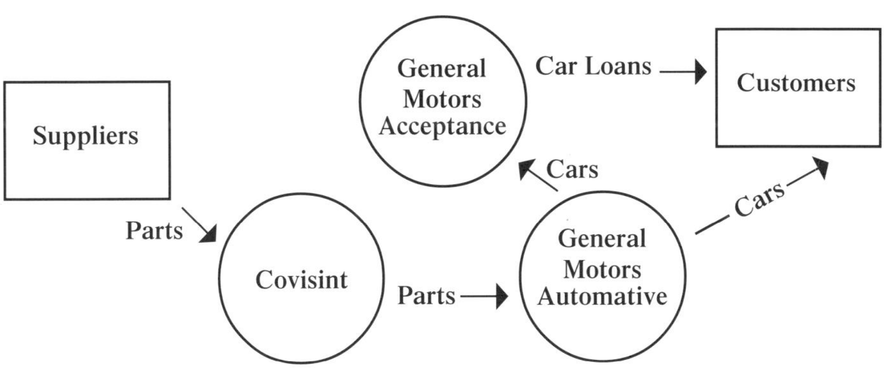

Chapters 4 and 5 described in detail the decisions required to draw the value map. The boundary boxes serve as devices to gather up our assumptions about markets – prices, capacity beyond which these markets will saturate (for consumer markets) or at which we will hit capacity (for vendor markets). They essentially act as "sanity checks" on our assumptions about the quantity of transaction flows. In drawing the Value Map, the analyst must identify ownership, influence and responsibility for each value flow. In the process, the economic events central to the value generating process are defined, as are the 'equivalent units' used to express them across all of the value map's processes.

> ***Note:*** Repeat steps 2 through 6 for each value flow (arrow). These steps are applied once each to each of the two components of value: (1) the average transaction value in a period (i.e., the transaction value); and (2) the count of transactions in a period (i.e., the transaction volume).

Step 2: The next step requires the analyst to explicitly define the influence of (1) technology choice and (2) managerial strategy on: (1) transaction value; and (2) transaction volume. This is not information that is necessarily the subject of accounting or audits, but it is available from press releases and annual reports. Management will also use press releases and annual reports to discuss the firm's technology and strategy, along with other information about the company. In addition, analysts can scan news releases and other information, including proprietary company sources. Technology companies like IBM, Cisco, Siemens, and others have their own monthly publications that tout their particular strategy and technology investments. Though much of the information is qualitative, it's quantitative implications can be surmised by building strategy models, and applying the scaling laws discussed in my previous book, *Valuing Technology*.

In Financial Dynamics, the *technology choice* portion of the *strategy model* defines how management's choices of particular technologies force them to make particular value trade-offs over time. At this point, we have identified four scaling 'laws' (refer to Chapters 4, 5 and 6 of *Valuing Technology* for descriptions of technology acceleration, intelligence scaling, geographical scaling and organizational scaling*)* that describe how management can expect future costs and revenues to scale with respect to specific value drivers. This need not be considered an exhaustive list. In the future, there is no doubt that other general characteristics of technology will be discovered; it is also reasonable to assume that there may be industry specific and firm specific technology scaling 'laws.' There is, no doubt, considerable research that that can yet be done in technology scaling.

In Financial Dynamics, the *strategy model* articulates mathematically the way in which the analyst thinks that management's strategic decisions will generate value (or, conversely, will destroy value). Strategy models are firm specific; management makes specific claims about the influence their particular decisions and actions will have in generating future value for the firm.

We use the historical values of technology and strategy drivers during the estimation window to 'reality check' the validity of our technology and strategy models. The outcome is not a binary 'valid / not valid' judgment; rather it shows up as more or less accurate forecasts. Where the explanatory power of the technology and strategy models is not particularly good, then the future forecasts of values will exhibit high variance, and thus will be highly uncertain.

Both *technology* and *strategy* models are deterministic – they are stated as functions of specific drivers (which I represent with X's in appendix B) and of value components – either transaction value or transaction volume (which I represent with Y's in appendix B). Financial Dynamics also invokes the assumption of *value additivity* – that each given strategy or technology choice will define its own model in the form $Y_i = f_i(X_1, \ldots X_n)$ where Y_i is the value generated by this strategy or technology choice, the X's are value drivers, and $f_i(X_1, \ldots X_n)$ is a mathematical function (the model) specific to the ith technology or strategy. All of these individual models can then be added to compute aggregate value $Y = \sum f_i(X_1, \ldots X_n)$.[4] Additionally, dynamic effects like technology acceleration will change Y, over time.

Step 3: Financial Dynamics adopts a statistical approach that predicates future forecasts on the historical series of values and quantities in the estimation window. The mathematics of forecasting are presented in Appendix B. Financial Dynamics maintains an emphasis on computing an *unbiased estimate of value*. The statistical forecast estimator presented in Appendix B was developed with *unbiasedness* as an objective, while *efficiency* and *consistency* have been traded off. Though other objectives might be entertained, this is consistent with current accounting practice which places emphasis on a point estimate of the account value.

Financial Dynamic's assumes that future value flow is an estimate of some future (and thus unknown) variable based on historical fact and its underlying factors. It assumes that all environmental factors outside of those predicted or controlled by management's technology and strategy choice will remain roughly as they have in the past (i.e. we make a *ceteris paribus* assumption unless otherwise stated). This approach contrasts with the existing 'categorize and count' valuation approaches common to accounting and finance. The latter lack model components required for consistent forecasting and for gauging uncertainty.

To compute forecasts, Financial Dynamics first inputs an 'estimation window' time series of *at least* 10 values for each of: (1) average transaction value; and (2) number of transactions. The forecast value flow (a random variable) is comprised of two components:

(1) the *transaction volume* (a random variable) in a year; and
(2) the *transaction value* (a random variable) for a single transaction.

Table 6.4 shows a typical set of input values in spreadsheet form

Table 6.4 An example of input for transaction value and volume (GM's annual value flow and volume figures, 1989–2001[23])

Year	Operating Income ($millions)	Quantity Sold
1989	20444	8019000
1990	15862	7901000
1991	12257	7310000
1992	13173	7373000
1993	17060	7851000
1994	21565	8379000
1995	25065	8320000
1996	20169	8381000
1997	23978	8597000
1998	21568	8149000
1999	26049	8786000
2000	25373	8083120
2001	25533	8083120

From the same estimation window, also input the *value drivers* – the particular quantitative or qualitative measures that are expected to driver future valuation. Note that value drivers are distinguished from transaction values and volumes. Entering value drivers for the estimation period makes it possible to 'reality check' management's assertions that these will really have the claimed impact on future value generation. They may either be a time period, as is the case with technology acceleration or geographical scaling; they may be market size, as is the case with network externalities; or they may be unique to particular managerial policy and strategies.

Step 4: Estimate the *value contribution* from the firm's *business environment* (θ in the notation in Appendix B) and the firm's *own discount rate* ($1 - \rho$ in the notation in Appendix B) for the time series of: (1) transaction value; and (2) transactions count. Once you have set up the technology and strategy models in additive form, then you can use the *non-linear least squares program* to calculate an estimate. You will need to do this twice – once for *value* and once for *quantity.*

Step 5: Predict (1) transaction values; and (2) transaction counts for a minimum of 20 years into the future using the formulas in Appendix B, and the planned values of strategy and technology drivers targeted by management or assessed by analysts. Step 5 will produce (1) a schedule of forecasts for transaction values and transaction counts; (2) estimates of the value contribution from the firm's business environment (θ in the notation in Appendix B) and the firm's own discount rate ($1 - \rho$ in the notation in Appendix B) for each of the two forecasted time series; and (3) an error term (ε in the notation in Appendix B) for which we will have the sample statistical distribution.

Step 6: The analyst armed with the results of step 5 has at least two choices for use of this data:

1. Compute the forecasted 'value flow per period' for each estimated future period by multiplying transaction value and count. If these are rendered as random variables by using the information in the error terms, then a convolution formula will need to be used. The resulting schedule of forecasts for 'value flow per period' then can be discounted back to the present using a weighted average of the

two estimates of the firm's own discount rate computer (i.e., one for the transaction value, and one for the transaction count)

2. Compute two present values: one for the transaction value, and one for the transaction count, using the respective own discount rates for each, then multiply these to come up with the present 'value flow per period.' If the forecasted values and counts are rendered as random variables by using the information in the error terms, then appropriate statistical formula will need to be used to compute the distribution of the discounts. It is recommended that numerical methods be used in this case, as the mathematics become relatively involved.

Step 7: The final step involves reporting the results in a manner that will be most meaningful to the users of the analysis. Reporting options will be discussed at length in Chapters 9 and 12.

APPENDIX B: STATISTICS OF THE FINANCIAL DYNAMICS' FORECASTING MODEL

Financial Dynamics forecasts a stream of future value flows in a dynamic context. The estimation model for either transaction count or transaction value is (where either is represented by Y in the forecasting model in this appendix)

$$Y_t = \sum_i f_i(X_{i,t}) + \theta + u_t(\rho) \; t = 1, \ldots, \mathrm{T}$$

where

- $\sum_i f_i(X_i, \ldots X_n)$ is the strategy model where the X are *strategy drivers* – i.e., the factors that are actively managed in the firms strategy and technology choice. As was described in step 2 above, Financial Dynamics invokes the assumption of *value additivity* – each given strategy or technology choice will define its own model in the $f_i(X_1, \ldots X_n)$ specific to the ith technology or strategy. All of these individual models can then be added to compute aggregate value $\sum_i f_i(X_1, \ldots X_n)$ where the 'hat' over the $\hat{Y}$ indicates that this is our estimate of the strategy effect on Y.

• θ represents other contemporaneous components of value not influenced or anticipated by managerial strategy and technology choice. Specifically, θ is estimated by $\hat{\theta}_T$ which solves the least squares minimization

$$\min_{\theta} \sum_{t=1}^{T} \left(Y_t - \sum_i f_i(X_{j,t}) - \hat{\theta}_T\right)^2$$

• ρ represents strategy 'lag,' i.e., the autocorrelation of current value generation with the prior period value creation, where the error terms u_t follow a first-order autoregressive model $u_t = \rho u_{t-1} + \varepsilon_t$ where autocorrelation $|\rho| < 1$ and the ε_t's are independent with zero means and independent of the $X_{i,t}$'s. An approximation of the firm's 'own discount rate' for this particular set of strategy and technology choices is provided by $1 - \rho$.

The estimate of θ is obtained by ordinary nonlinear least squares estimator $\hat{\theta}_T$ that solves $\min_\theta \sum_{t=1}^{T}\left(Y_t - \sum_i f_i(X_{j,t}) - \theta_T\right)^2$. Subsequently $\hat{\rho}_T$ is estimated by the empirical correlation of the estimated residuals $\hat{u}_t = Y_t - \sum_i f_i(X_{i,t}) - \hat{\theta}_T$ with the estimate of $\hat{\rho}$ being the first-order autocorrelation of the $\hat{u}_t$. The formula is:

$$\hat{\rho} = \frac{\sum_{t=1}^{N-1} (\hat{u}_t - \bar{u})(\hat{u}_{t+1} - \bar{u})}{\sum_{t=1}^{N} (\hat{u}_t - \bar{u})^2}$$

where the mean is $\bar{u} = \frac{\sum_{t=1}^{N} u_t}{N}$. Once these calculations are complete, the prediction of Y_{T+h} depends on the one period strategy lag $\hat{\rho}_T$, and the strategy driver values at $T+h$ as follows

$$\hat{Y}_{T+h} = (\hat{\rho}_T)^h \left(Y_T - \sum_i f_i(X_{i,T}) - \hat{\theta}_T\right) + \sum_i f_i(X_{i,T+h}) + \hat{\theta}_T$$

The schedule $\{\hat{Y}_{T+1}, \ldots, \hat{Y}_{T+h}\}$ of forecasted values *of either value or volume* are the unbiased forecasts of value generation. The net present value (NPV) of volume or value is computed as:

$$NPV = \sum_{h=1}^{H} \left(\frac{Y_{T+h}}{(2 - \rho)^h}\right)$$

1 The definition follows the approach of Malinvaud, E. *Lectures on Microeconomic Theory*, Amsterdam, North-Holland, 1985, Ch. 10, 264-314

2 On the value map, the value and volume on a particular value flow "arrow" are each random variables; and following the normal convention in statistics, this is indicated by placing a "tilde" ~ over the character. A value flow that is stated as a random variable has an expected value as well as a dispersion measure such as variance or standard deviation, which may be perceived as one way of depicting risk.

3 When Financial Dynamics is relying on existing financial accounting for its data inputs, the standard periods will generally be years.

4 I discussed value additivity in Chapter 8, p. 137 of Valuing Technology in connection with the improvement of adjusted present value (APV) approaches over discounted cash flow (DCF) approaches. This improvement comes from value additivity – i.e., robust models will split a project into pieces, value each piece, and then add them back up. Value additivity improves on DCF by applying the valuation model to each of a business' various kinds of value flow (breaking out the capital structure and tax implications associated with each cash flow) and then adding up the present values of each flow to compute the value. When value is stated in terms of cash flow, the approach is called adjusted present value (APV), also called valuation in parts, value integration and valuation by components. In APV the most fundamental split is along the line of 'real' cash flows – revenues, cash operating costs, and capital expenditures; and 'derivative' flows – i.e., those associated with the firm's financing program, values of interest tax shields, subsidized financing, issue costs, and hedges. By dividing the valuation up into additive sub-problems, it does away with the highly nonlinear impact of weighted average cost of capital (WACC) errors in the valuation. APV is inherently more robust, because any errors in assessing future capital structure or taxes are simply added together, rather than being magnified by the 'cost of capital' power function which discounts future values back to the present.

CHAPTER 7

Technology Statics

"Well, in our country," said Alice, still panting a little, "you'd generally get to somewhere else if you ran very fast for a long time as we've been doing."

"A slow sort of country!" said the Red Queen. "Now here, you see, it takes all the running you can do to keep in the same place. If you want to get somewhere else, you must run at least twice as fast as that!"

– Lewis Carroll,
Through the Looking Glass

STATICS AND DYNAMICS

Just as our valuation approaches need to keep track of stocks and flows of value, so does managerial strategy dichotomize into decisions that alter particular aspects of the status quo (statics) or that change the course of evolution of the business model (dynamics). Of course, individual decisions may, and often will, do both.

Advances in technology, and the resulting transition from industrial to knowledge economy, have created new roles for technology in managerial strategy. For example, the digital computer has metamorphosed several times, each time creating new roles for itself. In the beginning, it was merely a toy that received the derision of Thomas Watson Sr. when he stated that there was worldwide demand for "perhaps five computers." His son, Thomas Watson Jr., proved him wrong with the IBM 360 line of computers that automated corporate accounting (though offered no other strategic benefits than reducing accounting costs). The personal computer, in its modern incarnation, appeared on April Fool's

Day 1980 when Steve Jobs and Steve Wozniak rolled out the Apple II. The ensuing surge of information-processing power on the desktop empowered individuals and departments, engendered the rise of the securities salesman who could do his own analysis (used to great effect in the early 1980s by Michael Milken and his bond traders at Drexel Burnham), flattened firm hierarchies, and ultimately brought about the reengineering juggernaut in the early 1990s that, as much as anything else, was a revolt against the command-and-control centralization facilitated by the mainframe computer. The next metamorphosis saw the stand-alone PC attach to the Internet in the mid-1990s to become a smart network terminal, with the world's information resources at its call. Instant messaging, blogging (web-logging), and Internet-based supply chain management became *de rigueur* as individuals and firms networked globally. In the early 21st century, no serious strategy can ignore the costless networking that the Internet provides with customers and suppliers, developers and employees. And the computer continues its evolution. It is now so small and powerful that a variety of "appliance shells" can be built around it – cameras, telephones, notepads, personal robots. These will be the stuff of the next revolution in strategy.

The firm's "vision" and "strategy" define future policies, make assumptions about the future competitive environment, and define current investments (or sunk costs) and future direction of use of these investments to generate value. This chapter and the next define the route that Financial Dynamics takes to incorporate managerial strategy with respect to technology into the firm's valuation.

THE CLASSIC MODEL OF COMPETITIVE STRATEGY

The knowledge economy demands that we distance ourselves from the antiquated commodities trading business model underlying financial accounting. Before you know a firm's operating earnings, you have to understand its operations – its strategic objectives, its calculation procedures, its relationship to the market and to its competitors. The standard framework for this analysis was published in the early 1980s by Michael Porter, immortalized in his "five-forces" model (see Figure 7.1).

Figure 7.1 Five forces defining the competitive environment

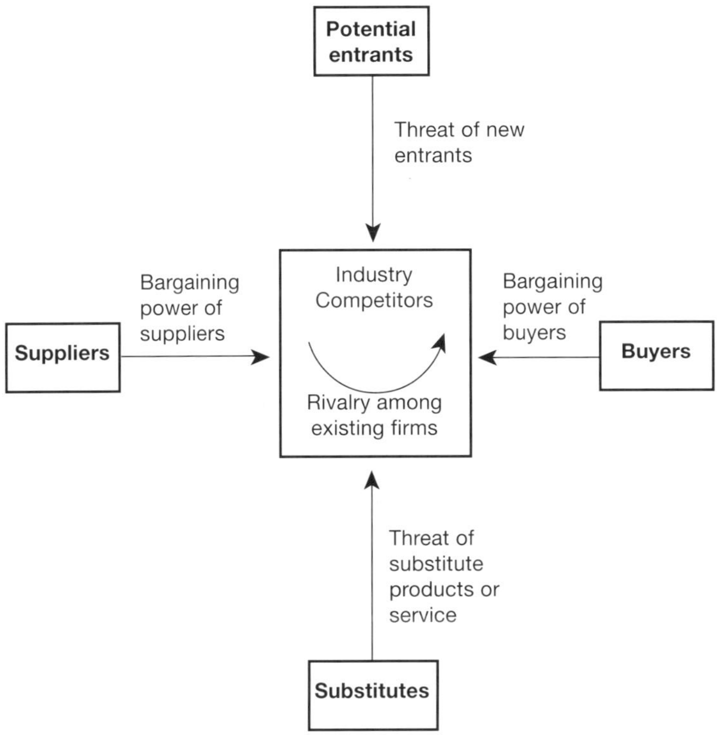

Source: Michael Porter, Competitive Strategy (New York: Free Press (1980)

Although Porter's five-forces model has become somewhat of a *de facto* standard for couching strategic decisions, it is important to remember that it was designed prior to the introduction of the personal computer. It is a product of our industrial era, and predates the dramatic impact on strategy of information processing on the desktop, empowerment, global reach, and instantaneous communication with customers and suppliers. This does not necessarily obviate its validity; rather, we need to be alert to technology-specific forces that may not have been considered by Porter at the time of creation of this model. One specific effect that has been studied extensively by Porter is "clustering," and this will be addressed in the next section.

In the five-forces model, *rivalry* is measured through indicators of industry concentration. The concentration ratio (CR) is one such measure. The Bureau of Census periodically reports the CR for major Standard Industrial Classifications (SICs). The CR is the percentage of market share owned by the largest *m* firms in an industry, where *m* is a specified number of firms, often four, but sometimes a larger or smaller number.

A high CR indicates that a high concentration of market share is held by the largest firms – that is, the industry is concentrated. With only a few firms holding a large market share, the competitive landscape is less competitive (closer to a monopoly). A low CR indicates that the industry is characterized by many rivals, none of which has a significant market share. These fragmented markets are said to be competitive. The CR is not the only available measure; the trend is to define industries in terms that convey more information than distribution of market share.

In contrast, the U.S. Department of Justice uses the Herfindahl-Hirschman Index (HHI) in guidelines for evaluating mergers. The HHI is calculated by taking the sum of the squares of the market shares of every firm in the industry. For example, if there were only one firm in the industry, it would have 100% market share and the HHI would be equal to 10,000 – the maximum possible value of the index. At the other extreme, if there were a very large number of firms competing, each of which had nearly zero market share, then the HHI would be close to zero, indicating nearly perfect competition.

Compared with many industries, the software industry is relatively unconcentrated. The total share of sales accounted for by the four largest firms in 2000 was just 26.7%.[1] Roughly half of all manufacturing industries have a four-firm concentration greater than that. More specifically, the HHI of industry concentration for software was 244 in 2000, versus 2,506 for automotive and 2,446 for breakfast cereals. Industries with an HHI of less than 1,000 are considered competitive by the U.S. Federal Trade Commission. By this measure, the software industry operates in a highly competitive market. As a practical matter, an HHI of under 2,000 is seldom a serious concern for the U.S. antitrust enforcement agencies.

Substitute products are products in other industries which apply different technologies (and thus cost structures) to compete in the same market as your firm. A product's price elasticity is affected by substitute products – as more substitutes become available, the demand becomes more elastic, since customers have more alternatives. A close substitute product constrains the ability of firms in an industry to raise prices.

Customers hold *buyer power* over a firm. By withholding or threatening to withhold purchases, they can exert control over pricing, profitability, and product quality. Where buyer power is significant, the industry is a monopsony – a market in which there are many suppliers and one buyer. Under such market conditions, the buyer sets the price.

The inputs to the firm – labor, components, and so forth – are acquired from suppliers, who influence firm strategy through *supplier power*. Suppliers, if powerful, can exert an influence on the producing industry, such as selling its raw materials at a high price and expropriating some of the industry's profits.

It is not only incumbent rivals that pose a threat to firms in an industry; the *possibility that new firms may enter the industry* influences competition. In theory, any firm should be able to enter and exit a market, and if free entry and exit exists, then profits always should be nominal. In reality, however, industries possess characteristics that protect high profit levels of firms in the market and inhibit additional rivals from entering the market. These are barriers to entry.

CLUSTERS

Clusters are geographic concentrations of interconnected companies, specialized suppliers, service providers, and associated institutions in a particular field that are present in a nation or region. Clusters arise because they increase the productivity with which companies can compete.

Michael Porter has initiated two studies of technology industry clustering in the United States. The *Cluster Mapping Project* is intended to put together a detailed picture of the location and performance of industries in the U.S., with a special focus on the linkages or externalities across industries that give rise to clusters. The second project, the *Cluster Meta-Study*, "aggregates data from existing cluster publications to learn more about the location, economic, and competitive characteristics of industry clusters, the reasons behind their competitiveness or uncompetitiveness, and their patterns of evolution over time as well as the reasons behind these patterns." The products of these studies should provide useful data for building specific Financial Dynamics strategy models for use in deciding whether or not to physically locate firm operations within a cluster.

At this stage, there do not appear to be similar studies on a global scale. Jeffrey Sachs of Harvard's Institute for International Development, has written extensively about the potential imbalances in global development which are implied by clustering.[1] He observes that since the end of the Cold War, the main factor dividing the world is not ideology, but technology. Specifically, 15% of the earth's population provides nearly all of the world's technology innovations (mainly from Japan, Europe, and the U.S.); a second part (50% of the world's population) is able to adopt these technologies in production and consumption (especially important in Asia); and the remaining part is technologically disconnected, neither innovating at home nor adopting foreign technologies. (This involves a huge swath through Central Asia, all of Africa, and South America.)

This imbalance has created huge problems for the developing world over the past two decades. Development in the industrial economy could take advantage of countries with cheap labor and scarce capital. The returns on new investments should be high, which, with proper governmental management, could promote saving and attract inflows of capital from abroad. This could be used toward developing an emergent middle class and a stable government. Asia has been the de facto leader in this scenario. In contrast, development in the technology-intensive knowledge economy allows regions with advanced technologies to best innovate further, since new ideas are typically produced from a recombination of existing ideas, and there are huge and growing learning curves for any modern technology. This favors "clusters" of innovation, critical masses in one place of linked industries and institutions, such as Silicon Valley and Hollywood. These clusters affect competition in three ways: (1) by increasing the productivity of companies based in the area; (2) by driving the direction and pace of innovation; and (3) by stimulating the formation of new businesses within the cluster. These clusters arise directly from the unique character of information assets. First, they are essentially public goods, and despite the best-laid plans, secrets tend to "leak" and the inventors of an idea tend quickly to lose control of it. Because ideas quickly become public goods, free markets are not enough to manage an ideas-based economy. The production of ideas – that is, innovation – may either be a product of basic scientific insight (based mainly on ideas in the public domain) or of applied engineering (backed by patents). The first relies on universities and public laboratories; the second on private, profit-driven firms. The problem for developing economies is that, when it comes to technology imports, countries must be successful as exporters of something to pay for the imports of technology innovations. This is the Red Queen's paradox thrown up by the knowledge economy. It takes all the running you can do to keep in the same place. If you want to get somewhere else, you must run at least twice as fast as that! A slow sort of country is quickly relegated to the sidelines of global competition.

BRAND EQUITY

Clustering has its main impact on research, development, and production of the firm's goods and services. Several information-specific effects occur on the demand side, in the consumer markets. One of the stickier issues in the valuation of intangible assets is that of putting a price on a brand, such as Dell, Firestone, or Sony. Though the valuation of brands is an exercise whose results are regularly divulged by business journals, we also know that these values are volatile. Two brands in recent history have seen an abrupt reversal of fortune – Firestone and Arthur Andersen.

Valuation of brand equity usually estimates some form of aggregate consumer utility (value). Since personal utilities are unobservable, this is naturally a highly subjective process. The theory is that brand equity is the difference in value (utility) of a product's features with and without the brand. This usually takes one of three forms: monetary, intangible, or perceived quality. Monetary value is the amount of additional income expected from a branded product over and above what might be expected from an identical, but unbranded product. For example, grocery stores frequently sell unbranded versions of name brand products; intangible value is the value of a product that cannot be accounted for by price or features. For example, Nike has created many intangible benefits for their athletic products by associating them with star athletes; perceived quality is the overall perception of quality and image attributed to a product, independent of its physical features. Mercedes and BMW have established their brand names as synonymous with high-quality, luxurious automobiles.

Only *perceived* quality seems consistent with the reversals at Firestone and Arthur Andersen. Quality, measured in terms of measurable aspects of the products or services, probably did not change at all for these two firms. But *perceived* quality did, because specific events – lawsuits for rollovers in the case of Firestone on Ford SUVs, and abuses at Enron for Andersen – provided information in an arena where understanding was murky. Tire technology is a black art that combines chemistry, physics, sewing, and cooking; it is fine-tuned on company tracks by racers who can't describe what makes a good tire, but know it when they ride it. Auditing is not well understood either, partially because of low public interest and a Byzantine rule set that confounds even its practitioners. It was designed in the 1930s when business was substantially different. As I will argue in Chapter 11, accounting firms' brand equity is overvalued. They are perceived as delivering a higher-quality service than they are capable of, and any public information that surfaces is likely to lower brand equity.

A *perceived quality* perspective on brand equity is supported by economic work in *adverse selection.* Adverse selection occurs in a competitive market where buyer and seller have different information about the quality of a product. The classic example is the "market for lemons," referring to George Akerlof's example of the used car market.[1] In his example, the demand for used cars vanishes as soon as *average* quality drops even though there are many good-quality cars. Obviously, the example is extreme. It is meant to make the point that the presence of unidentifiable "lemons" makes it difficult to sell cars of better quality at a good price, with the implication that the average quality of traded used cars is lower than that of non-traded used cars.

To make my point, let me ask you to contemplate, for a moment, the story of piano maker Yamaha. The company would successfully branch into snow skis, motorcycles, jet skis, tennis rackets, golf clubs, industrial unmanned helicopters, and machine tools, but originally was founded in 1887 by Torakusu Yamaha to make pianos. The company built up an enviable reputation for quality (it won prizes at the 1904 World's Fair for its pianos and organs), but suffered with the rest of Japanese industry during World War II. When president Genichi Kawakami took over Yamaha in 1950, it only made a few pianos, mouth-organs, and wind-up gramophones.

Brand was always an obsession with "emperor" Kawakami, as the press liked to call him. In 1954, to develop a market for pianos as well as the Yamaha brand, he started a music class for young children that evolved into the worldwide chain of Yamaha music schools (as well as showrooms for Yamaha's musical instruments). By the late 1980s, Yamaha had become the world's leading maker of musical instruments of all kinds, not just pianos. Yet, the ultimate brand recognition – the sort of prestige that went with, for example, a Steinway piano – eluded Yamaha. "A Steinway is a Steinway," said John Steinway. It was an arrogant pronouncement that struck a faultless chord with Steinway's public.

Acoustic tests and painstaking consumer trials assured Kawakami that Yamaha pianos sounded superior to Steinways. Indeed, Yamaha pianos earned a loyal following among jazz musicians for their superiority. But that mysterious quality – prestige – bedeviled many Japanese manufacturers who set out to conquer the world. Japanese cameras are the most technologically advanced available. Yet, for rich Japanese, Germany's Leica carries the real prestige. Yamaha motorcycles are known for their handling, reliability, and style. Who would prefer a finicky Ducati or cumbersome Harley-Davidson to a stylish, powerful Yamaha sportbike? Unfortunately for Yamaha, quite a lot of people.

Yamaha's painstaking attention to product quality and consumer preference has made its name, arguably, synonymous with quality, design leadership, and performance. Consumer magazines and product testing assure that the public is aware that Yamaha makes the best pianos. Yet, the exclusivity of a Steinway, Fazioli, or Bösendorfer has assured the latter brands a mystique that eludes Yamaha. It is obvious that branding is not *just* a way of signaling (that is, influencing *perceived quality*) when a particular firm has a better-quality product. There are also effects at the opposite, "luxury" extreme. These are called *Veblen effects* (a term for "conspicuous consumption" after Thorstein Veblen who wrote about them at the turn of the 20th century in *The Theory of the Leisure Class*). As a good becomes rarer or more exclusive, the price of the good rises, which some customers may see as a premium signaling greater quality.

The economics of perceived quality are well explored, and can provide us with strategy models for incorporating management's decisions on branding, target markets, and quality. Their application will be specialized, as they require a considerable amount of information about competitive and substitute products and their "perceived quality" in comparison with the firm's own products. Their application also requires an understanding of the demand curve for the product, which will be a complex function of quantity, price, and quality.[2] The data collection demands of such models are high, but are necessary where brand is a significant component of value, as it is in luxury companies such as Ferrari, Ducati, or Cartier.

NETWORK EXTERNALITIES

Network industries play a crucial role in modern life. The modern economy would be very much diminished without the transportation, communications, information, and railroad networks. Most of the major information industries are also either network industries or share many essential economic features with network industries.

Because of the nature of networks, the economics of a single networked member can rarely be analyzed in isolation. There are usually external factors – i.e., externalities – that influence the value of networked services. It is inherent in the structure of a network that many components of a network are required for the provision of a typical service. Thus, network components are complementary to each other, and network services demanded by consumers are composed of many complementary components. For example, interactive ordering while browsing through an Internet bookstore requires several components: a database engine to track inventory, transmission of signals, decoding through an interface, display on a computer monitor, and so forth. Clearly, there are close substitutes for each of these components; for example, transmission can be done through a cable TV line, a fixed telephone line, a wireless satellite, or PCN; but different combinations of various components will not result in identical services.

In *Valuing Technology* I alluded to emerging research on the value added by networks that in a broad sense support the intuition of Metcalfe's law – i.e., that the value added is proportional to the square of the number of users of a network. That value added only appears, though, after the number of users reaches a 'critical mass.' An individual needs to communicate with a critical mass of between 10% and 50% of the potential user base before that product will benefit from the network externality. This sort of squared growth of value appears when consumers make decisions independently of one another, and where the network itself does not actively manage the communication between consumers. It tends to appear in random networks, and those with regular linkages such as Bethe lattices. But such networks do not exhibit the 'small world' behavior so often found in real world networks.

Small separations are present in most real world networks. Species in food webs appear to be on average two links away from each other; molecules in the cell are separated by an average of three chemical reactions; scientists in different fields of science are separated by four to six co-authorship links; and the Web holds the prize at nineteen degrees of separation between arbitrarily selected Web pages.[1] At the turn of the 20th century, Guglielmo Marconi, Italian Nobelist and radio communications pioneer, asserted, based on his work with radio communications, that each person could be linked to every other person in the world through 5.83 hubs. This is perhaps the first articulation of the concept of 'six degrees of separation' or the 'small world' phenomenon in social networks. Empirical evidence was added in the 1960s by Manfred Kochen and Itheal de Solla Poole extrapolating from the empirical work of psychologist Stanley Milgrom. (All had met each other and collaborated, along with Benoit Mandelbrot, during their post-WWII employment at the University of Paris.)

Recent research suggests that assumptions that might lead to Metcalfesque behavior are violated in many networks that are important to us. Albert-László Barabási, a physicist at the University of Notre Dame, and his student Réka Albert have provided the most thorough summary to date.[3] They have shown that the Internet in particular, partly because it consists of commercial and not-for-profit components each managed with different objectives, exhibits very complex behavior. Indeed, they have suggested that we may be limited to understanding the Internet only in terms of emergent behavior, always in flux. The Internet is perhaps our most complex man-made invention to date, and is constantly being modified to meet one objective or another. Even if its development were static, the interactions between users would constantly alter its behavior.

Network externalities can be quite rare in practice. When they do occur, they are often quickly brought under control by government regulation. A recent example was brought to the public's attention on July 8, 2002, when eBay, the successful Internet auction site, acquired PayPal, the successful Internet payment service, for a US$1.5 billion stock transfer. For all its success, eBay had been slow to provide buyers with a simple, inexpensive way to pay sellers online. PayPal excels in this regard – 61% of its business comes from auctions, and it has become the payment scheme of choice for eBay. Users hold a PayPal account backed by a credit card of a bank account, and transfer money simply by typing in an email account. Both eBay and PayPal benefit from significant network externalities and a "virtuous circle," and both command more than half the business in their respective industries.

This raised the specter of government intervention, and in 2002 the acquisition was under scrutiny by the U.S. Department of Justice. There was also the possibility of eBay being regulated as a bank both in and outside the U.S. With eBay profitably serving 37 countries, it provided a very tempting target for regulation. The moral should be that once a firm becomes big enough to experience network externalities, there are likely to be other factors that blunt the potential for runaway growth. These may ultimately undermine the "virtuous circle" that was supposed to perpetuate this growth.

For those interested in knowing more about the current state of affairs in network research, Barabási, in his book *Linked*,[4] provides an enjoyable and accessible history of the science of networks. He effectively summarizes what we know about network behavior, and where this is likely to take us in the near future. This should be required reading for anyone interested in developing strategy models for network industries. Some simple networks, such as those encountered in peer-to-peer file sharing, may indeed behave as Metcalf predicted. But increasingly complex networks are displaying unexpected emergent behaviors that can yield entirely different economics.

---

1 Albert-László Barabási (2002) Linked: The New Science of Networks, Cambridge, MA: Perseus Publishing, p. 34 based on his estimate of around 800 million Web pages in 1998

2 John Guare, the author of "Six Degrees of Separation" is probably responsible for the promulgation of Marconi's 5.83 estimate. I think it is very likely that Marconi did offer this prediction to this degree of accuracy, because it is easy to see how he derived the number. If as Guare says, the estimate was stated in 1903, then it would have been a consequence of Marconi's experiments with transatlantic transmission. In 1901-2, Marconi was experimenting with a transmitter erected at Poldhu on the coast of Cornwall. His Nobel Lecture states that his two assistants on the *S.S. Philadelphia* were able to receive 'readable' messages up to 1551 miles, and test letters up to 2099 miles – transmission mileages stated to four significant digits. Assume he laid out a grid of squares, each with a transmitter in the centre, covering the inhabited area of the Earth. About 49 million square miles or about 15% of the Earth's surface is inhabited, and coverage of this area would require 5.83 transmission stations (rounded up to six, since the stations only come in whole units) each with a 1551 mile broadcast radius. Marconi was not referring to a social network he was just proposing how many strategically placed radio stations it would take to link all of Eath's inhabitants.

3 Albert, Réka and Albert-László Barabási (2002) Statistical mechanics of complex networks, Reviews of Modern Physics, v. 74, January 2002

4 Albert-László Barabási (2002) Linked: The New Science of Networks, Cambridge, MA: Perseus Publishing

CHAPTER 8

Technology Dynamics

Anything that won't sell, I don't want to invent
Its sale is proof of utility, and utility is success.
– Thomas Alva Edison

TRIBAL WARFARE

William Henry Gates III is the embodiment of Emerson's maxim that "an institution is the lengthened shadow of one man." From the moment Microsoft was founded, everything about it – brilliant, bad, or ugly – has been a pure crystalline reflection of Gates's mind, personality, and character. Before Gates arrived on the scene, the computer industry had always been organized vertically. Companies such as IBM and DEC would build their own machines, design and manufactured their own chips, and develop their own operating systems and applications – all of them proprietary. Gates envisioned a different structure, a horizontal structure, in which specialized competition would take place in each layer of the industry: chip company versus chip company, software company versus software company, computer company versus computer company. He held that the position of maximum power and profit in this new structure came from owning one of the critical industries – of controlling a "knowledge cluster" in a particular narrow product niche.

Michael Porter, in his book *On Competition*, emphasized that technology R&D likes to cluster. For example, there are clusters in regions such as Silicon Valley (semiconductors, software, electronics), Hollywood (film production), New York and London (finance), Santa Monica (special effects, post-production, and graphics software), Detroit (cars), Hsinchu (semiconductors, printed circuit boards), and on and on. Such clusters have sprouted up, in most cases unplanned, to take advantage of the unique economics of technology.

Microsoft's audacity was in threatening the Valley-centric industrial cluster from his boreal retreat in far-off Redmond. This, exacerbated by Gates' own quirky demeanor, left him alternately reviled and derided as an outsider in the industry, albeit one who was hard to ignore. Knowledge industries around the world tend to be controlled by clusters of bright people, working in geographical proximity. Microsoft challenged the *tribal status quo* with alternative ideas, prejudices, and lifestyles. And Gates won. Repeatedly.

Microsoft's troubles with the Department of Justice began in earnest in the spring of 1996, with the aspirations of two passionate, smart, and articulate Silicon Valley lawyers – Susan Creighton and Gary Reback, both antitrust specialists with the Valley's pre-eminent law firm, Wilson Sonsini Goodrich & Rosati. Reback's client roster included Silicon Valley's most prominent firms – Apple, Sun, Borland, Netscape, and Novell. That he was Redmond's most relentless and strident critic endeared him even more to his clients. He was known to have once remarked: "The only thing J. D. Rockefeller did that Bill Gates hasn't done, is use dynamite against his competitors!"[1] In Reback, Microsoft faced an adversary with both technical savvy and antitrust expertise.

The U.S. Department of Justice's antitrust case against Microsoft was a milestone in tribal wars wrought by the move from an industrial economy to a knowledge economy. Central to the case was Microsoft's so-called browser war with Netscape. Now that the dust has settled on the case, it is hard to say *what*, if anything was at issue in the dispute. In 1996, 75% of Web users used Netscape, which was based on knowledge that Marc Andreesen brought with him from the Mosaic browser developed at the National Center for Supercomputing Applications at the University of Illinois. No doubt, Microsoft used its expertise to catch up and compete head-on with Netscape with its Internet Explorer 3. (Any conspiracy theorist worth his salt will tell you that Microsoft always waits until version 3 to package the good stuff.) But it seemed a bit futile to charge Microsoft with "predatory pricing" in a contest in which both browsers were being given away for free. The fact is that these were intellectual products distributed at zero marginal cost, since they were downloaded from the Internet, but requiring substantial up-front development costs by their makers.

Despite the marketing one-upmanship of Netscape technology versus Microsoft technology, it is unlikely that either used much in the way of proprietary technologies. Both Microsoft and Netscape were conversant in the same technology, and shared both employees and information, some of it freely available to anyone through open-source software. In software, any new discoveries quickly percolate across corporate lines. In such an environment, not only can market leadership *not* be locked-in by clever technology, first-movers, or low prices, but the only way to succeed in such a market is through active management of the market for the technology, and through linking and leveraging with other products (precisely where Microsoft excels). Netscape never developed a business model to exploit these realities, and has since shriveled to a diminutive division within troubled AOL. This was simply one more advance in Gates' bid to force the Silicon Valley tribes to pledge fealty to Redmond gods.

The parable of Netscape and Internet Explorer illustrates two things about the knowledge economy. First, the production of knowledge begins and ends with bright people. And these people often stick together, living and working in the same geographical area, like latter-day tribes. Michael Porter and Jeffrey Sachs both refer to this phenomenon as "clustering." Second, these same bright people have trouble keeping a secret. Once produced, knowledge quickly "leaks" out to the rest of the world. In fact, in a perverse move to accelerate the leakage of software knowledge, bright folks in the software tribe have fostered the "open-source" community with the likes of Linux. (Take *that*, Bill Gates.)

The problems posed by "knowledge leakage" to the old ways of accounting are enormous. Our policing and legal systems, which define and protect property rights, have not at all kept up with business as operations and products that have grown increasingly knowledge-centric. The defense of intellectual property weighs in as one of the great social, political, and legal challenges posed by the knowledge economy. In truth, intellectual property requires a complete revision of economic systems and concepts of ownership.

To date, our methods of monitoring and policing intellectual property rights are for the most part ineffective. Consider just one example, that of the music industry. A series of file-sharing programs – Napster, Morpheus, Kazaa – have circumvented any of the methods developed by RIAA, Sony, or other parties to stop the widespread, unmonitored dissemination of music which, legally, belongs to someone.[2] No matter what our intentions at the societal level, this music has essentially become a public good. I for one, if I had to bet, would not predict that these parties will ever regain control over their songs and other copyrighted work. Not only that, but the file-sharing programs are also effective at distributing television shows, movies, and software globally the instant that they hit the market. Like it or not, Kazaa makes any intellectual asset a *public good* – like street lighting and highways – available to everyone in the world who is interested in using them.

Technology and other assets of knowledge-intensive businesses are similarly *public goods*. That does not mean that companies cannot exploit what property rights are available. For example, patents are designed to encourage innovation; to protect novel, useful, and non-obvious ideas. Yet, many firms are pursuing a more perverse route – patent hording. IBM obtained over 2,700 patents in 2000 (900 on software alone) and earned more than US$1 billion in revenue from intellectual property licensing. Oracle, Walker Digital, Lucent, Sony, and Motorola were up there too. Publisher Tim O'Reilly of O'Reilly & Associates observes: "Everyone knows that the system is broken, but they all choose to exploit it." There is no doubt, either, that the consuming public is less than sympathetic to these claims; views that are expressed forcefully in websites such as www.Boycott-RIAA.com.

Technology, knowledge, and information tend to be "leaky" in that whoever controls them is likely to lose control over them very quickly. The economics of technology is thus tied very much to substitutes and complements and to how quickly these improvements to the technology reach the marketplace over time. In *Valuing Technology,* I described several "scaling" effects that are specific to a given technology, and dependent on its being a de facto public good, with constant and accelerating improvements and evolution. In this chapter, we will look at the impact on value of the three of these that vary over time, at a constant, universal rate, across industries: (1) technology acceleration, (2) geographical scaling, and (3) organizational scaling. This is why we refer to these effects as "Technology Dynamics." All three arise from the "leaky" character of the knowledge that describes the technology, and because of this they are globally the same at any particular point in time (or nearly so; certainly close enough for the long-range forecasts required for valuation).

The only way in which these technological effects are firm-specific relates to what and how much of a technology or its substitutes are used at that specific firm, which in turn is a byproduct of managerial strategy. Thus, for specific technologies, global tables may be consulted (theoretically, though they are just now starting to be compiled) and applied to specific firms where appropriate.

It is important that these effects be explicitly incorporated into our strategy models. Existing tools of measurement and analysis ignore leaky information. This chapter presents arguments for particular functional forms of the strategy model component attributable to each of the dynamic effects. These functions may then be plugged into the Financial Dynamics strategy model to reflect the value impact over time of a strategy of investment in a particular technology.

STRATEGY IN TECHNOLOGY CHOICE

Strategy in technology firms has grown extremely complex over the past decade. Today's managers have to be much smarter than ever before to do it well. They also need new tools to decide how to extract value from technology investments. In technology-intensive business, the classical tools of valuation – double-entry bookkeeping and discounted cash flow – are inadequate, relying on business models from a simpler era, and produce results that may be inaccurate or irrelevant to current strategy.

The most striking change brought about by the increasing complexity of technology is the need for more and more investment not directly related to constructing the revenue-generating goods and services of the firm. In extractive industries such as mining and petroleum, or in low-technology production such as household consumables, expenses such as R&D, process control, quality assurance, customization, and so forth are minimal. These "support activities" become the major contributors to corporate expenses in knowledge-intensive technology companies. They dwarf the "primary activities" that are directly involved with assembling and packaging the products that are sold. Software represents an extreme – it is entirely comprised of fixed costs of supporting activities; the marginal cost of producing an additional copy of software is essentially zero, even though there is a positive unit revenue expected from each sale.

Strategy is a rational process of deliberate calculation and analysis, designed to maximize long-term value. By default, it is carried out in the market-spaces of equity capital, R&D, products, and services.

The majority of value in the knowledge economy is created by the application of strategy, ideas, and technology to products and services in advance of ever seeing any revenues. Branding, support, and research and development may be the major, if not the sole, expense – the ante, so to speak – associated with a profitable product. These costs are incurred prior to selling the first unit of a product. Consider Microsoft's X-box game console, designed and developed by Microsoft, but produced entirely by Flextronics, who also made Microsoft's keyboards and mice. The economics of the X-box, as far as Microsoft is concerned, are little different than if they owned a nanobox.

The steady rise in complexity of products and services had shifted value away from purely physical assets over the prior two decades. Daniel Bell, in *The Coming of Post-Industrial Society*,[3] marked 1971, after two years that resulted in a four-fold increase in world oil prices, as the beginning of post-industrial society, contrasting it with previous agrarian and industrial societies. In industrial society, the factors of production were often summarized in the mnemonic "manpower, money, machines, methods, and materials." Materials were the scarce and costly inputs to a production method (the knowledge component of industrial wealth creation). Money in the form of capital and revenues bought the necessary machines and labor to complete the transformation of raw materials into finished goods, generating wealth and more revenue in the process.

Between 1970 and 2000, the role of "manpower" altered dramatically. Skilled labor grew rapidly in importance, while, especially in developed economies, unskilled jobs migrated away to the country that could field the cheapest labor. In 1970, manpower substituted directly with machines (indeed, microeconomic models made such substitutability a standard assumption), and machines in an assembly line embodied the relatively static methods of production. Over time, methods themselves grew fluid and dynamic, embodying a rapid response to customer dissatisfaction and a changing competitive environment. Firms adopted names like Flextronics to advertise their competitive flexibility and responsiveness.

Today, over 80% of the wealth in developed countries is generated by information and service businesses – a reflection of the growing importance and complexity of methods and other knowledge assets. Intangible as they are, knowledge assets have grown to have enormous social, economic, and political importance. The ability to properly assess and value technology-intensive projects has become an essential component of global competitiveness, and needs to be embraced at the governmental as well as corporate level.

HOW TECHNOLOGY CREATES VALUE

Do technology investments pay for themselves? This, of course, is the central question asked in the so-called productivity paradox. The answer, in the aggregate, is positive. But many things can confound the choices available to the individual firm. For example, immediately after the dot-com bubble burst, U.S. productivity continued to improve, lending more support to the widespread perception that the dot-com bubble was more a financial scam than misrepresentation of the underlying productivity of technology. Output per man-hour in the non-farm business sector rose at an annual rate of 8.6% in the first quarter of 2002, its fastest growth in 19 years. Quarterly figures are volatile, yet the year-on-year growth in productivity was also impressive, at 4.2%. These numbers can be further broken down. The growth in output, and hence productivity, was inflated in the first quarter by a big swing in inventories, while firms actually continued to cut jobs. Productivity often surges in the first year of a recovery after recession, as firms produce more without needing to hire extra workers. The actual productivity growth is more likely to be around 2–2.5% – less than the 3–4% claimed at the height of the dot-com bubble, but above the 1.4% average over the two decades prior to 1995.

But there is a subsidiary question that is perhaps more important at the firm level. If technology does generate value, can this value be appropriated by the technology-adopting firm to improve its profitability?

Or is some of it likely to slip through, only to benefit customers, suppliers, and other third parties; or worse, simply to make the industry more competitive, less profitable, and more dependent on the technology vendors?

Unfortunately, recent evidence seems to indicate that the latter situation prevails.[4] Most of the benefits of America's productivity growth have not gone into companies' profits, but have been passed through to third parties. Since the 1950s, there has been a strong positive relationship between productivity and profits. But in the late 1990s, that relationship broke down. Despite a surge in productivity, national-accounts profits (as opposed to profits reported by companies, a less accurate measure) fell between 1997 and 2000 (Fig. 8.1), even before the economy dipped into recession.

Figure 8.1 Productivity (national-accounts profits) versus profits reported by companies

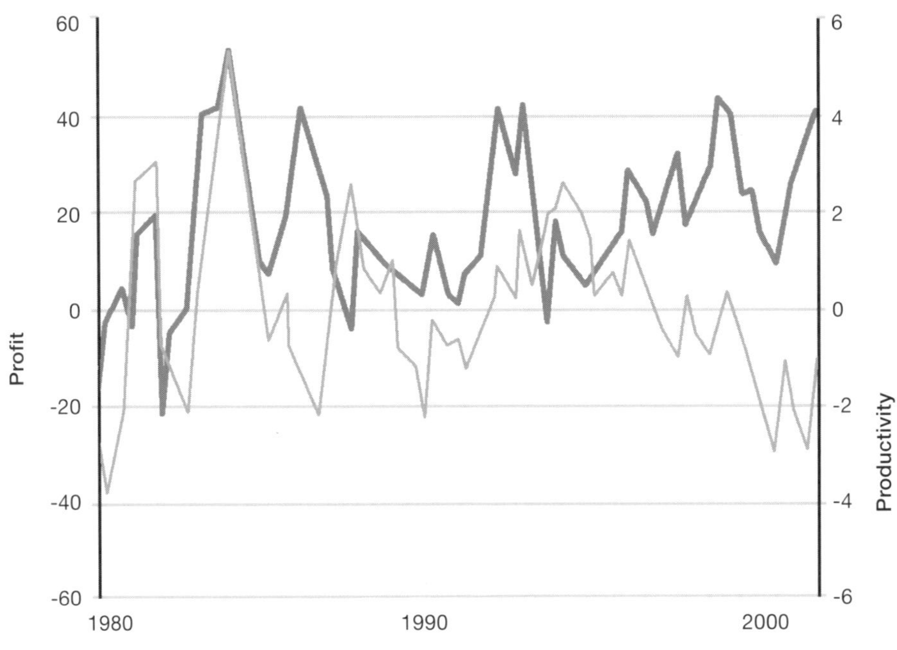

At the end of 2000, the profits of America's non-financial firms were no higher in real terms than in 1994, implying a big fall in their share of GDP. Workers (who are, naturally, also consumers) were virtually the sole beneficiaries of the new economy, in the shape of faster real wage growth. This was partly thanks to a fall in the prices of IT goods that they bought. More important, the same IT that spurred productivity also increased competition more widely across industries, from airlines and banking to insurance and cars, squeezing prices and profits. Information technology reduced barriers to entry, and made it easier for consumers to compare prices. This is nothing new. Productivity gains from previous technological revolutions, from railways and textiles to electricity and the car, have gone largely to consumers. Each time, a decline in the prices of goods and services has given a big boost to real incomes. Consumers gained from cheaper travel or clothes.

Unfortunately, for individual firms, profits, and even survival, were elusive. Firms' experience with new technologies has largely been that of grappling with creative destruction, innovation, and productivity – the raw material of the knowledge economy. These thrive where the sentiment prevails that all employees are created equal, that whatever your background, each of us has a positive role to play in the success of an idea. Strong-minded activists who love their profession and products are a hallmark of successful firms in the 21st century. Coherence is honored, while diversity is celebrated.

Successful knowledge economy cultures can be expected to thrive on diversity. No doubt in the 21st century there exist quasi-feudal regimes where strategy and government restrains broadcast and education in pursuit of social control. And these regimes can be very successful, at least while they mobilize an underemployed labor force. But in the process of maintaining control, they cut themselves off from the new. Failed societies revel in their "histories" – both real and imagined – and in this way discourage innovation and renewal. Such systems can shine for awhile, but are ultimately doomed by what the Chinese have called the "three generations curse" – successes of the current generation are frittered away by sons of the dynasty.

The knowledge economy brings with it its own social contract; one that promotes the exchange of ideas and relinquishment of thought control. With the free exchange of ideas comes an increased risk of disorder, and an assurance of change, rewarding some groups and possibly depriving others. Winnings in the casinos of technology come at the cost of a risk of change and a loss of certainty.

Steps toward incorporating the impact of the knowledge economy, and especially the impact of information processing, have been made over the past five years by both the U.S. Financial Accounting Standards Board and in the National Income and Product Accounts.[5]

Figure 8.2 Average annual growth of software sales (a), and actual sales (b) in constant dollars

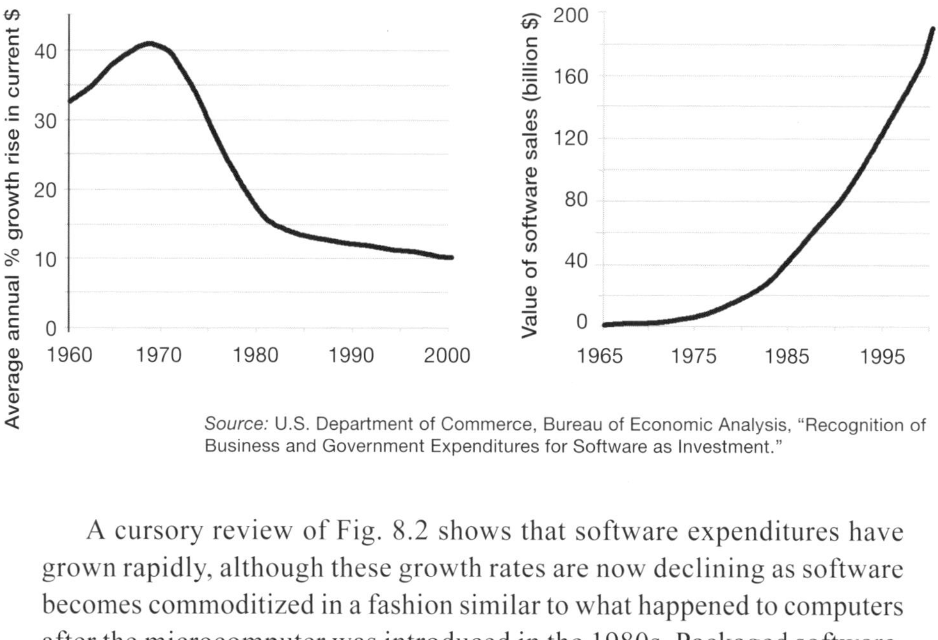

Source: U.S. Department of Commerce, Bureau of Economic Analysis, "Recognition of Business and Government Expenditures for Software as Investment."

A cursory review of Fig. 8.2 shows that software expenditures have grown rapidly, although these growth rates are now declining as software becomes commoditized in a fashion similar to what happened to computers after the microcomputer was introduced in the 1980s. Packaged software, shareware, open-source software, and other developments are putting downward pressure on revenues of software companies, but at the same time economies of scale allow many types of packaged software to remain profitable. Prior to the introduction of the IBM 360 in 1964, business and government procured only minuscule amounts of software – annual sales only reached US$1 billion in 1966.

Where software has a short service life – increasingly the case in packaged software, because upgrades make the old version technologically obsolete – the relative size of the software stock (that is, the asset value of software) is small in comparison to total investment. This trend has accelerated significantly over the past decade, as business has moved from custom mainframe software – which typically had a life of from 10 to 15 years – to packaged software, where service life may only be a few years.

Information technology has generally made industries more competitive, and profit margins have declined from an average of 16% in U.S. firms in the post-war period from 1948 to 1972, to around 11% today. It should be noted that a variety of other factors, including globalization, have had a significant impact on profitability as well. Information technology is more likely to impact personal wealth and productivity, and indeed it does. The growth rate of real GDP per person rose from under 2% in 1973 to around 3.5% in 2000; while the growth rate of real gross business product per person grew from around 1% in 1973 to about 2.8% in 2000.[6]

BUILDING THE STRATEGY MODEL

Financial Dynamics assumes that in the long run, value generation in the firm is only partly influenced by management's strategic choices; the remainder of value (or loss) generated is the result of factors inherent in the firm or its environment. These effects are explicitly segregated in the modeling. The process for building a strategy model is depicted in Fig. 8.3.

Financial Dynamics modeling is stochastic, but assumes that the strategy enacted within the firm is predicated on deterministic models of how that strategy will create value. These deterministic models may or may not accurately reflect the real world. Financial Dynamics reflects this by embedding management's policy model in a statistical model that:

- fits the policy model to firm outcomes through an analysis of historical data; and
- employs the fitted model to predict future value flows.

The predicted value flows may be retained in raw form – a form that is useful for scenario analysis, financial planning, and other time-sensitive tasks – or it can be compacted into a single "valuation" number using a discount rate. Reporting options appropriate to particular applications are discussed later in the book.

Financial Dynamics improves on existing valuation approaches by providing *formal roles* for all of the major classes of objective and subjective information at the analyst's disposal. Without specific formal roles for available information, analysts are driven to *ad hocracy* – that is, willy-nilly incorporation of information in incomplete or flawed fashion into the limited models available for computing value.

Figure 8.3 The roles of each type of information in assessing the value metric

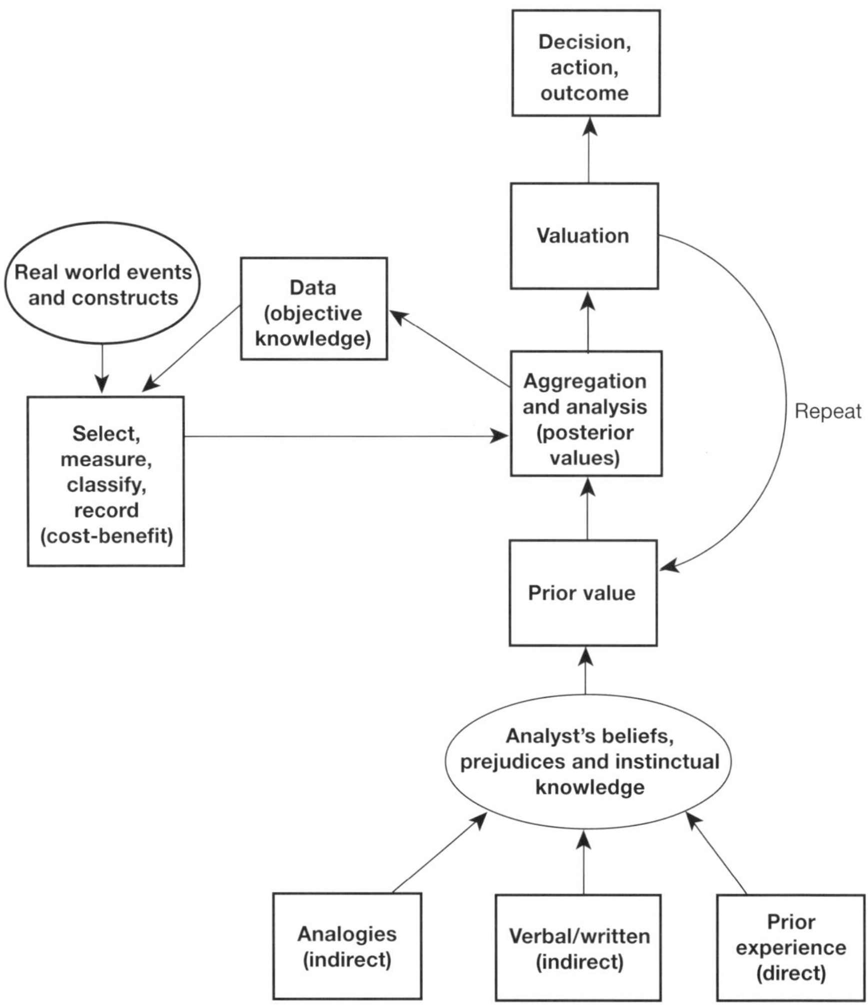

The classical example is the failure of GAAP accounting to report anything at all of risk, variance, or unit volume statistics in a transaction flow, let alone investor and consumer sentiment about the company and its products. Thus, forecasts (and financial valuations using forecasted future cash flows and earnings) are likely to exclude most of the information needed for accuracy. Analysts introduce these after application of DCF or other valuation approaches, and then in an *ad hoc* and usually imprecise way. Such forecasts may or may not be inaccurate. But they are likely to be inconsistent, and their assumptions, accuracy, transparency, and validity are difficult to ascertain – in fact, they are probably severely lacking.

Figure 8.4 describes the components of the Financial Dynamics method. Each of these will be described, in turn, in this chapter.

Figure 8.4 Structure of the Financial Dynamics method

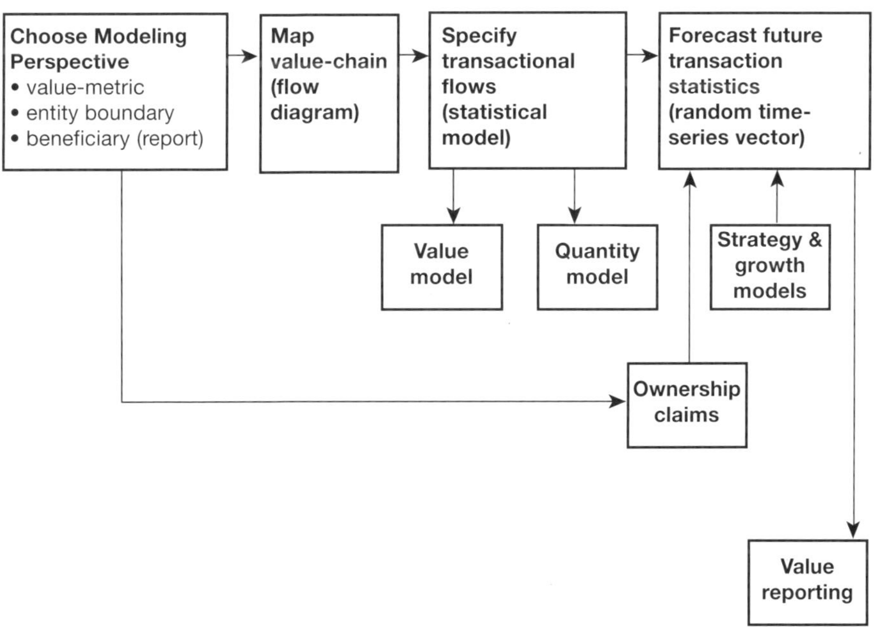

Financial Dynamics turns the perspective of traditional accounting for wealth accumulation on its head. In the agricultural and industrial economies, the value of a business was measured by its stored assets minus claims to those assets. This is implicit in traditional accounting reports that depict a business in vaguely articulated formal and informal processes that generate *stored wealth* (that is, assets) while using up, via expenses, some part of the firm's wealth endowment.

Financial dynamics focuses, instead, on the *processes* that impel the *value flows* through the firm. This makes much more sense in the knowledge economy, where the major (knowledge) assets are difficult to store, quickly grow obsolete, and, because they are "non-rival," may be impossible to unequivocally own or protect from theft. In the knowledge economy, value comes from "ideas in action." Knowledge assets possess negligible value until they are put to work.

In accounting's "stored wealth" perspective on corporate value, stocks have played the central role in audit and bookkeeping. In Financial Dynamics, stocks, if they exist at all, are just another cost of doing business. And usually *less is more*, whether we are speaking of inventory, equipment or any other traditional store of wealth.

The Financial Dynamics perspective is a "value flow" perspective; thus, the valuation model focuses on transaction flows that reflect the creation of value. These give us a time-series model that can both "learn" – that is, integrate new objective data – and also make the most of insights, qualitative knowledge, and other sorts of "subjective" data.

Often a firm's intermediate processes are difficult to value because there is no market for their outputs, or markets that exist are illiquid and erratically priced. In knowledge economy firms, where core processes are important and other activities are likely to be outsourced, critical processes will cross firm boundaries, alliances and joint ventures can be the predominant providers of value, and internal accounting tells almost none of the story. (Just consider how marketable Microsoft's products would be without the existence of computer and chip manufacturers.) Prematurely assigning dollar valuations to these processes robs the valuation of forecasting power.

Figure 8.5 How to choose a value metric

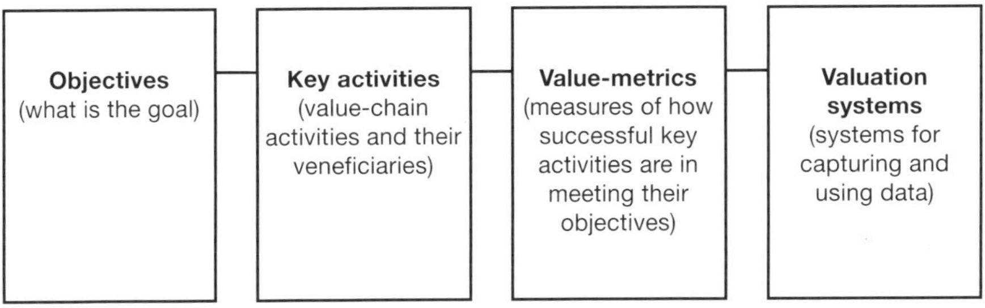

The *value measurements* are the specific outcomes of assessing the value of a given *metric* – that is, the *metric* is the conceptual "measuring stick" (Fig. 8.5); the *measurement* is the value that is recorded when this "measuring stick" is applied to a given event or transaction.

Technology-specific choices drive value creation in the strategy model. Thus, the particular parameters which management chooses, either explicitly or implicitly in their strategy, are called "strategy drivers" in Financial Dynamics' strategy model. *Technology acceleration* (a dynamic effect) is the improvement over time in performance for a given price of any particular technology. Network externalities (which are static effects) are increases in demand price with increase in market size for a particular technology. Organization scaling is a decrease in the cost of production resulting from the substitution of people with technology. And geographical scaling is the reduction in transport and communications costs resulting from improvements in transport logistics and communications switching infrastructure.

Most technology of importance today is available simultaneously to all firms competing in a given market. This is the result of two trends: (1) complex technologies are information-intensive, and that information easily disseminates throughout an industry as employees move, as academic papers are published, and through the exchange of ideas in professional societies; and (2) technology-specific tasks are increasingly outsourced to specialist firms, fragmenting production, and improving efficiency through improved market communication and logistics. If the technology creates real value, then there are significant competitive pressures to adopt it by all firms in a given product market. Technology adoption almost never provides sustainable competitive advantage. Rather, it periodically sets a new baseline for competition among all firms in an industry. Information technology is often a cost of doing business to the firm, with efficiency improvements passed on to the consumer in the form of better products at lower prices.

TECHNOLOGY ACCELERATION

"Technology acceleration" is the term I have used to describe the economy-wide ratio of performance to price of technology sold in the marketplace. This ratio increases – that is, it accelerates – over time as a result of the social, economic, and laboratory processes behind technology advances. Technology acceleration is related to the broader economic study of what are called *hedonic pricing methods*, which themselves are approaches to identifying *shadow values*. The hedonic pricing literature attempts to infer demand for product characteristics (for example, performance) from market prices. Automobiles, property, and houses are common subjects of such studies. By observing market data, we can infer the marginal value (price) of each characteristic from the associated partial derivatives. For example, the price of a car reflects the characteristics of that car – transportation, comfort, style, luxury, fuel economy, etc. Therefore, we can value the individual characteristics of a car or other good by looking at how the price people are willing to pay for it changes when the characteristics change. In contrast, technology acceleration describes how the price of that specific characteristic we identify as "performance" varies over time.

Table 8.1 shows data on technology acceleration in the motorcycle industry over the past 50 years, where the performance metric is the ratio of brake horsepower (bhp) to weight in kilograms (kg) for a set of motorcycles that have sold for around $8000 in real dollars c. 2000. The formula for value as a function of time is bhp/kg = $.207e^{.0272t}$ where t = *year -1949*. In this case, the technology acceleration coefficient is .0272 (which implies a relatively slow evolution compared to computers).[1]

Table 8.1 Technology Acceleration in the Motorcycle Industry, 1950-2001

Year	Model	bhp/kg
1950	Vincent Black Shadow	0.24
1973	Kawasaki Z1	0.36
1981	Suzuki GSX1100S Katana	0.46
1983	Kawasaki GPZ750 Turbo	0.48
1984	Kawasaki GPZ900R	0.48
1986	Suzuki GSXR1100	0.63
1991	Yamaha FZR1000	0.63
1992	Honda CBR900RR Fireblade	0.66
1998	Yamaha YZF-R1	0.85
2001	Suzuki GSXR1000K	0.94

It should not be surprising that technology acceleration is exponential. We are considering a change in perceived utility or productivity of a technology over time. It is common for our physical perceptions to respond logarithmically – for example, our eyes and ears perceive intensity of light or sound logarithmically, which gives their perception a range of scales that is hard to duplicate in machines. Similarly, our human perceptions of quality or other factors of human or social importance are likely to be logarithmic, and for the same reasons. If this is the case, then our perception of price for a given quality level – that is, the hedonic price – is likely to grow exponentially over time, simply because of the way that we are measuring/perceiving that particular quality. In the case of Moores' law, whether or not performance is measured in gigahertz or in Specmarks, we will obtain an exponential relationship. The unique character of human measurement is probably the basis of many of Gilder's "laws of the telecosm" and of what I have more generally called technology acceleration.

What can be perplexing is the tendency of technology acceleration to span different platforms – for example, in computers, Moore's law has held over mechanical, vacuum tube, transistor, and integrated circuit technologies. But recognize that in a competitive market, technologies compete in an environment where potential substitutes surround them (one of Michael Porter's five competitive forces). Whatever is the ratio of performance to cost , if a substitute improves dramatically enough, it becomes the lead technology contender. It's Darwinian, and it's not necessarily possible to predict which technologies will prevail. But it is possible to predict how quickly technological performance in pursuit of a particular market will improve. This is captured in specific values of the technology acceleration coefficient. Any particular technology acceleration coefficient reflect the choice of "performance metric" used to assess advancement and the time period over which advancement is measured (Table 8.2 shows the technology acceleration coefficients for a number of computer technologies, where we use commonly advertised performance metrics to compute the coefficient). In the example of Moore's law, the metric is computational cycles per second; in the example of Gilder's law, the metric is switching cycles per second. As with other metrics presented in this book, an annual time interval is assumed (as much to be consistent with tradition as anything else).

The archetypal example of technology acceleration was first proposed by Gordon Moore, co-founder of Intel, and is now eponymously called Moore's law. Moore's law predicts that semiconductor computing performance (measured as MIPS/cost) doubles every 18 months – that is, $P_t \propto e^{.462t}$ where the technology acceleration coefficient is .462. Similarly, Gilder's law becomes $P_t \propto e^{1.465t}$. The coefficients .462 MIPS/cost and 1.465 switching cycles per second/cost, respectively,[7] are annualized technology acceleration coefficients for the specific performance metrics – MIPS/cost and switching cycles per second/cost, respectively. Technology acceleration coefficients tend to be both unique and stable, a fact that will be useful in forecasting future technology acceleration. Moore's and Gilder's laws have held since the 1930s independent of a particular product or platform. This suggests that they are primarily demand-driven, and reflect the way that innovation responds to market demand.

Technology Acceleration in Hardware

The strategy driver in technology acceleration is *time*. The general functional specification of technology acceleration is $\frac{X}{P} \propto e^{\alpha t}$ where X is the performance metric, P is the price for a unit of performance, $\propto$ means 'is proportional to', α is technology acceleration, and t is the elapsed time. Thus time is the value driver; and for the technology defined by α, the price at time t is $P_t \propto \frac{X}{e^{\alpha t}}$

Table 8.2 Empirical values of technology acceleration

Technology	Acceleration* coefficient	Number of weeks to double	Performance metric
CPU	0.55	74 (~1.4 years)	Clock speed (in MHz)
Telecommunications switching	1.47	28 (~7 months)	Clock speed (in MHz)
CD-ROM	1.59	26 (~0.5 years)	Rotation speed
Printers	1.98	21 (~5 months)	Resolution * printing speed (ppm)
Non-volatile RAM	1.66	25 (~0.5 years)	Storage amount
Volatile RAM	1.47	28 (~7 months)	Storage amount
Monitors	0.23	179 (~3.4 years)	Size * resolution * frequency/point size

Source: Eric See-To HKUST Working Paper, September 2002

Technology Acceleration in Software

Though it is harder to come up with unequivocal "performance" or "quality" adjustments for purely knowledge-based products, such as software, there are an increasing number of metrics, usually from hedonic pricing studies.[8] These can be used to compute technology acceleration in intangible products. Since 1997, the U.S. Bureau of Labor Statistics has published a Consumer Price Index for "computer software and accessories" which indicates that the real price for a particular performance level in software fell about 6% per year in real dollars. This implies a software technology acceleration coefficient of $0.058 = \ln(1.06)$ $\Rightarrow P_t \propto e^{0.058t}$. This is considerably slower acceleration than semiconductors, but by historical standards is still rapid acceleration of technology development.

* Please note that the technology acceleration coefficients computed in my previous book *Valuing Technology* are all understated by a factor of 2.30. This resulted from my applying a log_{10} rather than a log_e function to the data. This and other errata appear on the book Website.

Many things go into the creation of computer software before there is any code. The creation of a software package follows some conception of the purpose of the package – for example, to check spelling in documents, to determine whether a number is prime, or to control a particular piece of hardware. Going from the conception to a workable package requires the producers to develop architecture for the software that will guide programmers in the coding process. It may also require special numerical algorithms or programming tricks. As a result, a software package may depend on certain intellectual property, protected by patents or trade secrets that are independent of the code that is actually written. For example, the popular MP3 audio format is based on patented algorithms held by Fraunhofer IIS.

Two distinctions are useful for summarizing the major types of software. The first is the distinction between distributing software as source code (open source) or as binary code (proprietary). The second is the distinction between distributing software at no direct charge (unpaid) versus for a direct charge (paid).

The spread of personal computers in homes and offices created a mass market for software. Microsoft initiated the concept of writing software for profit – a model that is now much imitated. These companies initially protected their investments by enforcing copyrights over their software and by distributing the software in binary form to prevent reverse engineering. In the mid-1980s, it became possible to obtain patents on certain aspects of software and many companies chose this route for protecting their intellectual property. The number of software patents awarded annually in the U.S. has risen from 829 in 1986 to 7,398 in 2000. Most proprietary software is sold at a price, and the source code is typically not made available. At the high end of microcomputer software is Mathematica, which is licensed to commercial users for $1,500–$2,500. The middle ground is held by applications software such as MS-Office, priced in the hundreds of dollars. Utilities are typically priced around $50, with "freeware" and "shareware" covering the bottom end of the range from being free to a few dollars. Open-source software distributes source code as well as binary code and is sold under either the older BSD license (Berkeley Software Development, which was primarily Unix-related development), which allows use of the source code as long as there is acknowledgment of the copyright; or GPL (General Public

License), which requires the source code for any modifications to be released under GPL. These provide the bases for "viral" models.

Worldwide, commercial software generated $171 billion revenues in 2000, up from $35 billion in 1988 for around 12% annual growth. In 2000, 45% of the sales revenue came from software written for Wintel (Windows on Intel) computer platforms. On all platforms, 23% of sales came from development software (for example, language compilers for C, or environments such as VisualAge); 45% derived from applications (for example, MS-Office); and the remaining 32% came from systems software and infrastructure (for example, Unix operating system, or Oracle database management system). In the U.S. alone, it was over $90 billion (almost all from sales of proprietary software) generated by around 10,000 companies employing around quarter of a million people, implying on average a small scale of firm operations.[9] In comparison, the motion picture industry was smaller at $74 billion and the auto industry larger at $427 billion.

The evidence from the successful software firms is even more impressive. Despite the NASDAQ's poor performance, business spending on software is about 75% higher in 2002 than it was in early 1995; Intel and Sun Microsystems had doubled their 1995 sales by 2002; overall S&P's information technology index rose at an annual rate of 11% from 1995 to 2002; inflation-adjusted wages are up 13% over these seven years, and productivity growth has sustained itself at 3.1% – far faster than the 1.5% annual growth rate from 1980 to 1995.

ORGANIZATIONAL SCALING

A Parable from Industrial America

The bursting of the dot-com bubble has been used by some as an argument that such technologies are specious, the products of exaggerated claims and market manipulation. But such bubbles are not uncommon when a new technology dramatically enters the economy. They result from intense uncertainty about the ultimate form the technology will take. Nowhere was this more apparent than in the automotive industry at the turn of the 20th century, and in the career of Henry Ford, the inventor who perhaps most embodied the American spirit of entrepreneurship and industry in that day.

The automotive industry evolved in the center of a "knowledge cluster." Detroit had grown to become America's premier motor and machine tool center over the 19th century as a result of the Great Lakes shipping industry. Ships laden with iron ore from Duluth would travel through the St. Lawrence Seaway delivering it to smelters in Pittsburgh (next to Pennsylvania's coal fields). They stopped halfway at Detroit, where a burgeoning industry built up around steam, then diesel engines on the ships. The development of "portable" versions of these engines was a natural move in Detroit.

Nearly 300 different cars were made and marketed in 1908 (often little different from each other); in 1909, 18 new firms began building cars; in 1910, 18 went bankrupt. Financiers worried about the exposure, as most were run by technicians rather than businessmen. By 1914, there were only 50 auto companies; by 1925, the year that Walter Chrysler started his company, over half of the cars sold were either Ford or GM; and the bursting of the stock-market bubble in October 1929 winnowed the automotive field to these three.

Henry Ford's technology was not radically different from that of his 300 competitors, and he started late as well (not unlike Bill Gates and Microsoft). But what catapulted him from relative unknown to kingpin was his defiance of the Selden Cartel. In 1895, George Selden, a Rochester, New York lawyer who had never built a car, applied for and received U.S. patent number 549,160 for an internal combustion engine. The patent was purchased by a consumer watchdog group, "The Selden Patent Group," which operated as a trust, similar to the Edison and Eastman cartels trying to regulate the newborn cinema. It used the patent to collect royalties, and to decide who could build cars, and how many they could build if permission was granted. Cantankerous Midwest farm boy that he was, Henry Ford refused to pay their royalties, or to let the Selden Cartel regulate their production, prompting a lawsuit in 1903 which Ford eventually won in 1910. The trial and his widely reviled adversaries gave Ford a public relations coup that made him an immediate folk hero.

The Model T was not only a better car in 1915 than it was in 1908; it also sold at half of the original price. There was no obvious increase in material, labor, or overhead invested in each car (in fact, the latter two obviously dropped). Instead, Ford's systemization and automation of production followed the principles of Fred Taylor's scientific management.[10] Fred "Speedy" Taylor, father of the stopwatch-and-clipboard approach to factory management, showed Ford and his managers during the 1913 year the time and motion techniques that had been implemented with so much success by Isaac Singer (sewing machines), Cyrus McCormick (reapers), and Samuel Colt (firearms). "Speedy" Taylor's *ideas* about the production of cars *replaced materials, labor, and overhead* that had previously been required to produce a Model T. In Ford's "Taylorized" Highland Park plant, one man could now do what three or four had done before. This knowledge leaked off to his competitors in no time. Within 10 months, Willie Durant (again with the help of Taylor) adapted the assembly line to assembling Chevrolets in Flint.

A Ford Model T (originally painted in red, green, blue, and gray varnishes, but in black after Henry Ford discovered in 1914 a superior Japanese paint which only came in black) cost $850 (a teacher's salary) in 1908. It came without speedometer, windshield wipers, or even doors, and the gas gauge was a long thin stick that the owner had to find for himself and insert into the tank. By 1915, assembly lines had allowed Henry Ford to incorporate speedometers, wipers, and doors and still lower the price to $440. It was down to $295 by 1925.

"Taylorized" mass production made farm boy Henry Ford so rich that, in 1914, embarrassed by his riches, he took 20% off the retail price of a Model T. When the result was even more sales, he announced that if 300,000 Americans bought Model Ts in 1914 he would return $50 to every buyer – a gesture that ultimately cost him $15 million. Next he raised the pay of his 13,000 workers from $2 per day to $5 per day. The raise was so steep (GM paid around $2.35 per day) that the *New York Evening Post* exclaimed it "a magnificent act of generosity." In contrast, the *Wall street Journal* accused Ford of "economic blunders if not crimes" by injecting "spiritual principles into a field where they do not belong."[11]

Much of the move toward mass production in those days involved standardizing parts and processes – in other words, *making transactions routine!* Alfred Sloan, the legendary boss of GM, learned one of his first lessons in efficiency as a GM supplier at Sloan's Hyatt Company, which made roller bearings. The dressing down came from Henry Leland, founder of Cadillac, for the tolerances he allowed in his roller bearings. With calipers in hand, Leland showed Sloan the discrepancies in bearing diameters, commenting: "You must grind your bearings even though you make thousands, the first and the last should be precisely alike."[12] Leland walked Sloan over to a window, showed him a pile of rejected Weston-Mott axles, and told him that unless he could guarantee bearings ground to within a thousandth of an inch in accuracy, the Hyatt products would also end up on the scarp heap. Only precision to within a thousandth of an inch made parts interchangeable.

Our parable describes many ways that the early automotive industry parallels other more recent technologies. The industry geographically grew up in a "knowledge cluster" which drove the pace, and provided investment for the new auto companies. The U.S. patent system was used to stifle creativity. The company that prevailed didn't necessarily possess better technology, but benefited from a combination of good marketing, good technology, sound production, and luck. Much of that success resulted from replacing materials, labor, and machines with knowledge. The benefits from the technology accrued mainly to customers (in better quality and lower prices) and workers (in higher wages); they made the industry more competitive and less profitable. The final shakeout in the industry involved the bursting of the stock-market bubble in 1929, with the ensuing economic despair and recriminations. Efficiency improvement through automation was largely a procedure of standardizing parts and processes – in other words, *making transactions routine.*

Substitution between Labor and Technology

Organization scaling is the raison d'être for investment in the firm's "back-office" information technology systems, factory automation, and other technologies for production efficiency. These systems increase the productivity of the firm's employees. Organizational scaling results from a *substitution* of technology for human effort. This is not a "1-to-1 man replaced by robot" sort of substitution; rather, a more subtle "three people can now do the task formerly allocated to four" sort. This substitution effect reflects the change in relative demand for labor or technology when their prices change, but the company's budget is held relatively constant.

Organizational scaling is a hybrid of technology acceleration and its influence on the *ad hoc* value metric chosen as the objective of firm activities – in publicly traded firms, this is typically profit, or earnings per share (EPS). Any particular technology is presumed to have its own unique acceleration coefficient that describes how performance/price will increase over time. In turn, that performance impacts the profitability of the firm (assuming the firm is maximizing profitability). The simplest general way to model this impact is through a Cobb-Douglas production function.

In knowledge intensive companies, labor grows more expensive because of greater skill requirements and wage inflation; while technology cost declines exponentially as a result of technology acceleration. The company's consumption of technology and labor in support of a particular process is constrained by the firm's budget B for the process of interest. Assume the process of interest uses X_T units of technology and X_L units of labor; or alternately, one can consider X_L to be the number of units of the numéraire (money) or more generally, the opportunity to spend on something else. Assume that the relevant transfer prices equal the marginal rates of substitution of X_T and X_L, call them (P_L, P_T) where $x_L P_L + x_T P_T \leq B$ and B is the budget for enabling and running the process. If this trade-off in consumption by the firm is depicted using a Cobb-Douglas function $x_T^{\,a} P_L^{\,1-a}$ then we can assume that the firm optimally spends aB on technology, and $(1 - a) B$ on labor, where $0 \leq a \leq 1$. The optimal choices for consumption of technology and labor in the Cobb-Douglas context are $x_L = \dfrac{(1 - a) B}{P_L}$ and $x_T = \dfrac{aB}{P_T}$ which implies that $a = \dfrac{P_T x_T}{B}$.

Assuming that the technology in question is available to all competitors in the field then P_T is just the market price for a particular unit of performance; x_T is the firm's consumption of the technology. Technology acceleration influences the *potential* for cost-effectively improving the efficiency of people with machines. This potential, though, is too often not realized, implying that the firm is not operating at the optimum. Failure to use technology efficiently may occur through bad management, demands for job security and labor inflexibility, or difficulty in integrating the technology into existing systems. Whichever cause is to blame, the result is the firm adopts an internal price P_T that is above the market price. In predicting future value flows, this estimate of efficiency is an essential part of ascertaining the organizational scaling.

We have considered performance-price ratio $\dfrac{x_T}{P_T}$ previously in our discussion of technology acceleration. But in this discussion, the ratio specified the performance available in the marketplace for the technology

– what has been called 'innovation diffusion' on the supply-side of the market. There are a variety of reasons that a given technology may not be adopted inside the company as quickly as it becomes available in the marketplace – unwillingness of the labor force to alter work habits or embrace the technology; lack of technology specific expertise in the firm; high cost of adoption; and so forth. As a consequence, *internal technology adoption* will accelerate slower than *external technology innovation*. If α is the external technology acceleration coefficient, then assume that for a specific firm or project, there will be an internal technology acceleration coefficient $\alpha' \leq \alpha$ and therefore $\frac{x_T}{P_T} \propto e^{\alpha' t} \leq e^{\alpha t}$. Using the *internal technology adoption* α' the budget constraint becomes $x_L P_L + K e^{\alpha' t} P_T^2 \leq B$ or alternatively $x_L P_L + \frac{x_T^2}{K e^{\alpha' t}} \leq B$ – choose one or the other depending on whether quantity x_T or acquisition price P_T of the technology is constrained in running the process.

We showed that an estimate of α,call it $\hat{\alpha}$, can be estimated by regressing published performance and price statistics to fit $\alpha t = lnx_T$ - lnP_T - lnK. Internal technology adoption acceleration $\alpha' \leq \alpha$ could theoretically be estimated using the firm's internal performance and historical cost statistics. But these statistics would not be readily available to external analysts, and in this case, α' could be estimated to be some value less than α based on judgments about the efficiency of the firm's technology adoption. Large, conservative, hidebound firms might be expected to have $\alpha' \cong 0$, whereas agile firms that constantly assess and reinvent themselves might be able to realize a value $\alpha' \cong \alpha$. Since the cost of technology will be changing over time, a given expenditure on inputs will yield increasing output, all other things being equal, because of the technology acceleration.

Consider the evolution of the early car industry, as reflected in the example of Ford's Model T. The automobile's technology acceleration coefficient was around $\alpha \cong .032$. During the period from 1908 to 1925, the price of Ford's labor was stable, with a 2.5 times leap in 1915 from \$2 per day to \$5 per day for 13,000 man-years annually, and output of around 300,000 Model T's annually in 1915. Assuming Ford was operating with a narrow profit margin, total budget was probably around $P_T x_T$, or in 1915 about \$440 × 300,000 = \$132,000,000 (or a more accurate estimate

could be obtained from the cost of goods sold on the firm's income statement). Assuming that the results are constrained by a demand of around 300,000 cars per year, this gives figures:
$\$5 \times 13000 + P_T Ke^{.032t}$ \$132,000,000 with $K = 660$ giving a cost of around \$440 per unit in 1915 (t is stated in years since 1915). Then the cost in subsequent years, as a result of technology acceleration, would be predicted as shown in Table 8.2.

Table 8.3 Predicted cost of a Model T, 1915–28

t	P_T
1915	\$440
1916	\$419
1917	\$398
1918	\$379
1919	\$360
1920	\$343
1921	\$326
1922	\$310
1923	\$295
1924	\$281
1925	\$267
1926	\$254
1927	\$241
1928	\$230

Marx, Marshall, and Organizational Scaling

The economist who had perhaps the most to say about the marriage of machine and man was Karl Marx.[13] Marx referred to the *ratio of labor to machines* as the "*organic composition of capital*." In Marx's mind, machinery could not produce value itself, but was necessary in order to gain a temporary price advantage over competitors. His reasoning: machines are a tool of man, and essentially an amplifier of their labor.

Marx published in the mid-19th century, just as the Industrial Revolution was gathering steam. Materials, labor, and capital represented distinct political groups in class-conscious Europe – materials from mines and farms of the landowning nobles (~3% of the population); capital from the savings and investment of bankers, merchants, and entrepreneurs (~10% of the population); and labor from the working classes, often poor, displaced serfs or sharecroppers (the remainder). These political forces were relatively permanent and introverted in the time before mass media. Their particular claims were legitimized in the account classifications used by accountants and economists alike.

In Marx's system, materials, labor, capital, and whatever else had to be valued in labor terms. This created conceptual problems for accounting and pricing – both central considerations in Marx's work. It took the brilliant and influential Alfred Marshall[14] to unravel these conundrums through a detailed analysis of price and demand behavior. Work prior to Marshall had usually presumed that there is a fixed supply of resources in the economy, and the only problem is how to allocate them among alternative uses. This was the assertion of the Physiocrats, who so influenced early accounting; William Beaver has shown that this "stewardship" orientation of accounting still dominates the objectives promulgated by the U.S. Financial Accounting Standards Board.[15, 16] Marshall argued that this is not the case. Resources grow through time, as does the population and the accumulation of wealth, technological progress, and a host of other economic influences. The ever-changing dynamics of an economy must be incorporated in policy formation and model-building.

Marshall is generally given credit for the original idea behind a "residual income" measure of corporate value – an approach that is at the core of valuation alternatives such as Stern Stewart's EVA.[17] Alfred Sloan implemented a residual income measure for General Motors' operating divisions in the 1920s, as did Matsushita in the 1930s and General Electric in the 1950s.[18] Residual income r in time period t is simply earnings minus cost of capital $r \times \kappa$ in time period t. This is $\rho_t = \varphi_t - [r_t \times \kappa_{t-1}]$ where ρ = residual income; φ = net operating income; κ = total invested capital; r = cost of equity capital; and t is the time period. I discussed this formula at length in chapter 8 of my book *Valuing Technology* including mathematical variations commonly applied in interpreting information from financial reporting. If residual income has an Achilles' heel, it is the need to find an acceptable cost of equity capital r. Normal practice invokes the capital asset pricing model (despite its controversial flaws) to compute a return which is extrapolated into the future in an *ad hoc* fashion. But the formula for residual income $\rho_t = \varphi_t - [r_t \times \kappa_{t-1}]$ is a mix of flows (ρ_t, φ_t) and stocks (κ_{t-1}), and r_t is simply the conversion factor which takes stock of capital κ_{t-1} and converts it into an outflow of resources (a cost) associated with the processes needed to calculate ρ_t.

Consider the steps involved in practical calculation of the cost of capital $r_t \times \kappa_{t-1}$. First a flow of individual expenditure transactions at time t τ_t is dichotomized into assets κ_t and expenses $\tau_t - \kappa_t$ and by deciding whether or not they purchased anything that will generate value for more than a year into the future. The sum of accumulated assets less depreciation, i.e., $\sum_t \left(\kappa_t - \delta_t\right)$, is then called 'capital' and is presumed to generate value at some 'rate' r_t. Recall that earlier we showed that discounting involves the fairly subtle choice between present and future consumption. – i.e., the stock of a good that you'd give up today for a promised stock of one unit of the same good one period in the future. In practice, this is usually set to some externally dictated 'objective' measure such as a US Treasury Bond rate – i.e., the number of 'treasury bond dollars' that you'd give up today for a promised 'treasury bond dollar' one period in the future. If you figure that all of these numbers are predictions of the future, and each has errors, misrepresentations and inherent uncertainty, you can easily see how the compounding of errors swamps any useful information out of the result. It is not that the concepts are necessarily wrong, but rather that the resulting calculations are so error-infested that they become useless. Wouldn't it be simpler to state everything in terms of flows to begin with?

Other Considerations in Organizational Scaling

The tradeoff between technology and labor has been studied extensively in the economics literature. In keeping with the general theme of Financial Dynamics, the approach I adopt here is the simplest one I could find that depicts the relevant policy components and the way that they influence value. Organizational scaling, though, is the most complex of our various scaling laws. As a consequence, I want to address two issues that may weaken the explanatory power of the Cobb-Douglas model. The first issue is highly variable budgets; the second is the desire for leisure.

The first problem arises when budgets are allowed to vary considerably. This typically occurs when a firm receives injections of capital from the stock market, banks or venture capitalists to fund a business segment. In this case the Cobb-Douglas functions assumption of a fixed proportion spent on technology and labor, *a* and *1-a*, will be invalidated. In this case the model can decompose the income and price effects using the *Slutsky equation*, after the Russian statistician who published it in 1915. Suppose that the value-generating process we are interested in uses x_L units of labor and x_T units of technology at prices (P_L, P_T) when the company budgets B per year for the process. Then the demand function for technology may be written

$$\frac{\partial x_T}{\partial P_T} = \left(\frac{\partial x_T}{\partial p_T}\right)_{B=B(x_T, x_L)} - x_T \frac{\partial x_T}{\partial B}$$

where the term $B = B(x_T, x_L)$ implies that the budget in this term is fixed at a point that keeps x_T, x_L constant. This is a mathematical way of saying that change in technology demand due to performance-price change due to technology acceleration is the sum of a price effect and an income effect. Each of these three terms is in theory observable.

The other effect involves knowledge workers demand for leisure. This has been observed, at one time or another, in most high-tech industries, where employees often want to take sabbaticals, change careers, demand flexible work schedules, or require other access to 'leisure time' as a part of their employment package. Quite a number of successful technology entrepreneurs have left work for a life of leisure, or at least what they consider to be more meaningful pursuits. As wages rise, individuals will increase the supply of work – but in a complex fashion due to 'income' effects. When wages rise, workers can consume more without increasing the amount they work. One of the things they can consume is more leisure, bought by working less. At a high enough wage rate, the income effect overcomes the substitution effect and workers will prefer to spend part of the higher wage on leisure. This causes the supply curve to bend backward (Fig. 8.6).

Figure 8.6 Backward Bending Supply Curve

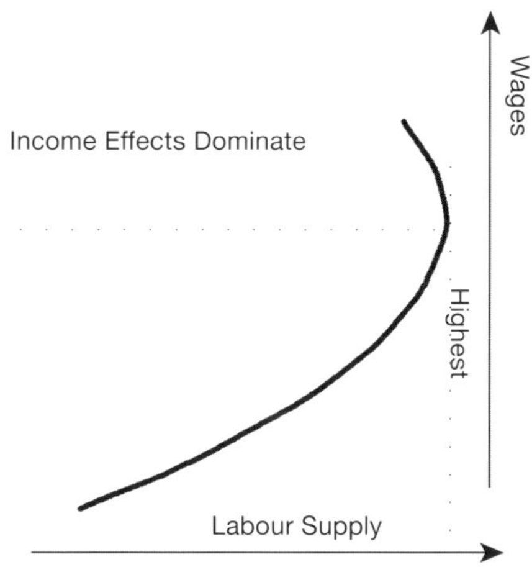

Technology acceleration, by making labor more productive over time, essentially shifts the labor supply curve up and to the right by an amount that accelerates over time, though in practice it is nearly impossible to predict when the curve will slope backward. The resulting change in behavior of the workers may also be difficult to predict.

Executive Pay

Labor in the time of Marx was a tangible political force, and the icon of labor sweating in the factories inspired many a Communist Party cadre under the leadership of Stalin and Mao. But in an era of clean rooms and microcomputers, sweating is discouraged by air conditioning, and workers might rather enjoy profit-sharing and flex-time than political clout. Worse, factory labor has declined to a minuscule proportion of the workforce – around 10% in the U.S. Executives, consultants, researchers, and a variety of other knowledge workers are the mainstays of 21st-century business. Gradation of wages up and down the hierarchy is hotly debated, as is the actual contribution to the firm at each level. In fact, the whole issue of hierarchy has been attacked with new vigor. Strategy guru Peter Drucker predicted that the knowledge economy would engender a corporate structure similar to a symphony (a group of equals under a conductor) or the British civil service in India.[19] The reengineering fad of the late 1980s helped many firms move toward these predictions. For example, one GM boss estimated that if a recently hired college graduate entered General Motors in the 1980s at the lowest rung in the hierarchy and spent two years at each successive rung, it would take *123 years* to move up to CEO. By shrinking the hierarchy, auto firms were dramatically able to improve productivity, allowing them to meet the Japanese challenge in the 1990s.

As a matter of basic fairness, Plato posited that no one in a community should earn more than five times the wages of the ordinary worker. Drucker has long warned that the growing pay gap between CEOs and workers could threaten the very credibility of leadership. He argued in the mid-1980s that no leader should earn more than 20 times the company's lowest-paid employee. His reasoning: If the CEO took too large a share of the rewards, it would make a mockery of the contributions of all the other employees in a successful organization. After massive increases in compensation, Drucker's suggested standard looks quaint. CEOs of large corporations in 2001 made 411 times as much as the average factory worker. In the past decade, as rank-and-file wages increased 36%, CEO pay climbed 340%, to US$11 million. "It's just way off the charts," says Jennifer Ladd, a shareholder who is fighting for lower executive pay at companies in her portfolio. "A certain amount of wealth is ridiculous after a while."

Oddly enough, CEOs came to command such vast wealth through the abuse of a financial instrument once viewed as a symbol of enlightened governance: the humble stock option. Throughout the 1990s, governance experts applauded the use of options, maintaining that they would give executives a big payday only when shareholders profited. And for a while, as the bull market ran its course, that's the way it worked. But as the market collapsed during the past two years, a funny thing happened: shareholders lost their shirts, but executives went right on raking in the dough.

GEOGRAPHICAL SCALING

Measuring Distance

The degree of geographical scaling experienced by a firm very much depends on the particular markets in which it operates, and the transaction volumes that flow between those markets.[20] Transport of physical goods has always been a significant part of world business – around 10–15% of gross world product. But as the world substitutes outsourced production with robotics, and replaces traditional physical goods with information technology (as, for example, has happened in the watch and camera industries), communications cost will increasingly determine the degree of geographical scaling. What physical transport remains is likely to shift toward high-value, minuscule size and weight semiconductors and electronics. Figure 8.7 gives a rough idea of the rate at which these changes have been taking place globally since the 1950s.

Figure 8.7 Semiconductors, electronics, and gross world product

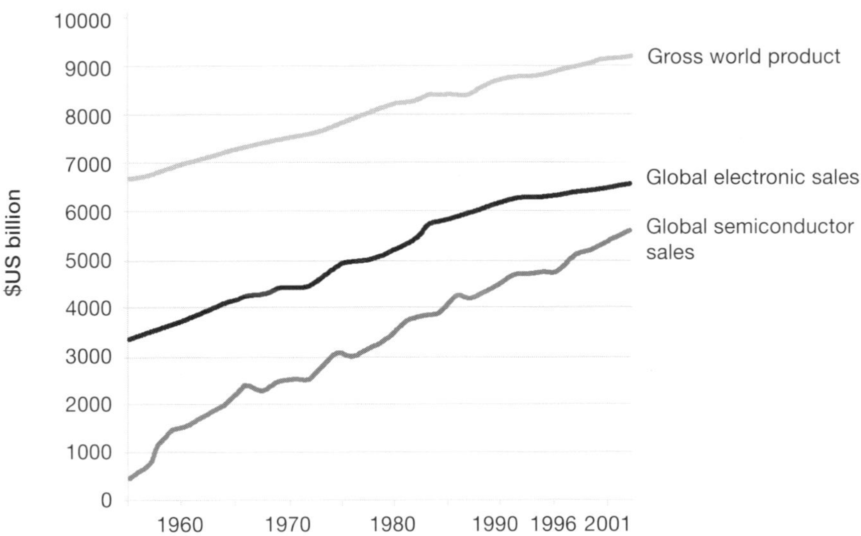

"Logistic" distance may be thought of in terms of the *cost* to transport a given package, defined in terms of size and weight, from point A to point B. This cost itself will depend on the distance, customs, and logistics facilities at points A and B, and scale economies – it is cheaper to send cargo on routes that are heavily used. To illustrate, it is cheaper to ship a machine from Los Angeles to Hong Kong, than it is to ship it from Los Angeles to Rosarita, Mexico, despite a several magnitude difference in geographical distance.

"Telecommunications" distance can be similarly thought of in terms of the *cost* to communicate a given package, defined in terms of size and

required security, from point A to point B. This cost itself will depend on the distance, communications facilities at points A and B, and required bandwidth. The implications for developing economies are obvious. Parts of India have become an attractive place for software development "factories," but a company in India competing for business from, say, the U.K. faces phone bills that are six times the cost of those from France and four times those from Israel, another country building international software competitiveness.

The European Union's European Telework Development (ETD) has recently compiled a measure of telecommunications "distance" based on the relative cost of international telephone calls (Table 8.4) [21]. "Distance" is shown relative to the cost of consumer calls between the U.K. and the U.S. (currently the most competitive international route), which is one unit of distance. National strategies in the U.S. and U.K. have made the links between these two economies the cheapest in the world. Other distances are shown as a multiple of this. For example, New Zealand is twice as far from the "centre" of the networked economy as the U.S. or U.K., Bulgaria is five times as far, Kazakhstan 10 times as far, and Afghanistan 15 times. It costs twice as much to phone from the U.K. to France (the UK's geographic neighbor) as to phone across the Atlantic. Even Australia, on the other side of the world, is closer than France in terms of telecommunications cost.

Table 8.4 Selected telecommunications cost-based distances

Country	Distance	Country	Distance	Country	Distance
United States	1	Hungary	3.8	Saudi Arabia	10.3
United Kingdom	1	Czech Republic	4.4	Korea - North	12.4
Sweden	1.3	Bulgaria	5.4	Kuwait	12.4
Canada	1.4	Brazil	5.9	Pakistan	12.8
Australia	1.6	Antarctica	6.1	Laos	13.1
Netherlands	1.6	Greenland	6.5	Nigeria	13.1
France	1.8	Tunisia	6.9	Sri Lanka	14
New Zealand	2	Colombia	7	Iraq	15.6
Ireland	2.3	Malawi	7.8	Afghanistan	16.4
Hong Kong	2.4	Uruguay	9.4	Eritrea	18.9
Korea - South	3	Kazakhstan	10.3	Cook Island	19.6
Malaysia	3.6				

Source: European Telework Development

Geographical scaling is the combined effect of changes in "logistic distance" and "telecommunications distance" over time. It is the raison d'être for logistics, supply map automation, electronic commerce, and many other applications of information technology in a business setting. These systems increase the velocity of transaction flows.

Unfortunately, there has been little work done in actually characterizing how quickly the world has shrunk over time, or on the substitution between 'logistics' and 'telecommunications.' In such situations, and until better information results from research, linear models tend to be most robust, as well as making the best use of limited information. Their results also tend to be easier to use. Thus I will suggest a functional form $value = k_{GS} \cdot t$ where k_{GS} is the geographical scaling constant, and t is the elapsed time over which we are looking at the changes in costs. By looking at specific histories relevant to our business model it is possible to come up with reasonable estimates for k_{GS}.

Consider the example of a trip from New York to Boston (Fig. 8.8). In the early 19th century, the gradual replacement of footpaths with stagecoach routes for business communications and transportation speeded up both travel and communications, since all communication was tied to newsprint, letters or other paper media. The telegraph, developed in the 1840s provided a substitute for communications, while stagecoach was still used for travel. This was eventually superceded by railroads, and then highways. Where a particular transportation method was not cost effective, it was abandoned.

Figure 8.8 The shrinking geography of New York to Boston travel (travel duration in minutes)

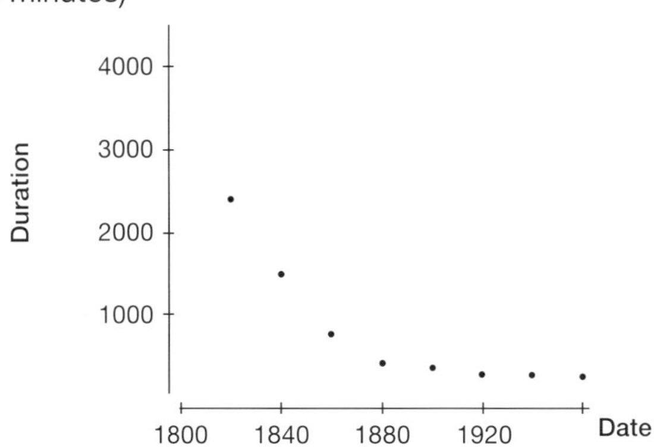

The data for New York to Boston travel suggests that the linear model may not be the best fit, thus we are justified in doing a bit more research. But as a first approximation, we can use 'duration' (the y-axis) as a surrogate for cost, and say that over a century, travel from New York to Boston drops to 20% of its initial cost. Annualizing, and in terms of our model that the value generated over t years is $value = .002 \times I \times t$, where I is the initial cost of the travel. Obviously this is very rough; but specific analyses will likely encourage the acquisition of better information about the evolution of logistics and telecommunications costs, which are imperative for the accurate valuation of technology firms.

[1] John Heilemann "The Truth, The Whole Truth, and Nothing But The Truth, "*Wired*, November 2000, p. 68–127.

[2] There is no doubt, either, that the consuming public's sympathy lies with the file-sharing companies. This is increasingly expressed forcefully at websites such as www.Boycott-RIAA.com.

[3] D. Bell, *The Coming of the Post-Industrial Society, New York:* Basic Books Reissue edition (May 1999)

[4] "To These, the Spoils," *The Economist*, May 9, 2002.

[5] Descriptions of changes in National Income and Product Accounts are summarized in E. Seskin, "Improved Estimates of the NIPA's for 1959–98: Results of the Comprehensive Revision," *Survey of Current Business*, 79, December 1999, p. 15–39; and in J. Christopher Westland, "The Cost of Errors in Software Development: Evidence from Industry," *Journal of Systems and Software*, 62(1), 2002, p. 1–9.

[6] This is discussed in J. S. Landefeld and B. M. Raumeni, "Measuring the New Economy," *Survey of Current Business*, March 2001, p. 23–40.

[7] Larger values imply faster acceleration; thus, a value of 1.47 for switching technology implies that bandwidth is accelerating three times as fast toward becoming "free" as is computing technology.

[8] Several studies have computed such relationships for software. See S. Oliner and D. Sichel, "Computers and Output Growth Revisited," *Brookings Papers on Economic Activity*, Vol. 2 (Washington, D.C.: 1994), pp. 299–301; E. Brynjolfsson and C. F. Kemerer, "Network Externalities in Microcomputer Software: An Econometric Analysis of the Spreadsheet Market," *Management Science*, 42, 1996, p. 1627–47; N. Gandal, "Hedonic Price Indexes for Spreadsheets and an Empirical Test for Network Externalities," *Rand Journal of Economics*, 25(1), Spring 1994, p. 164–70; and for a discussion of the construction of quality-adjusted price indices using hedonic methods, see Roseanne Cole, et al., "Quality Adjusted Price Indexes for Computer Processors and Selected Peripheral Equipment," *Survey of Current Business*, 66, January 1986, p. 41–50.

[9] IDC Report #25569, *Worldwide Software Market Forecast Summary, 2001–2005*, September 2001.

[10] Despite his popularity as a "business guru," Taylor continued to provide such services at his customary fee of US$35 per day.

[11] A. Madsen, *The Deal Maker* (New York: Wiley, 1999).

[12] Ibid., p. 92.

[13] Karl Marx has perhaps been unfairly ignored as an economic thinker (especially in the finance and accounting communities) because of his association with the dictatorial sort of Communism that arose in the Soviet Union. But Soviet Communism was brought about by abduction of the movement by the Anarchists within the Communist Party, and by Lenin, desperate to gain power in post-Tsarist Russia. The horrid publicity that ensued ironically caused Marx to exclaim on several occasions, "I am not a Marxist!".

[14] A. Marshall, *Principles of Economics* (1890, reprint by Prometheus Books, 1997).

[15] Financial Accounting Standards Board, *Statement of Financial Accounting Concept No. 1: Objectives of Financial Reporting by Business Enterprises* (Stanford, CA: November 1978).

[16] William H Beaver,. *Financial Reporting: An Accounting Revolution* (Englewood Cliffs, N.J.: Prentice-Hall, 1981); William H. Beaver and W. R. Landsman, *Incremental Information Content of Statement 33 Disclosures*(Stamford, CT: FASB, 1983).

[17] For example, see G. Biddle, R. Bowen, and J. Wallace, "Does EVA Beat Earnings? Evidence on Associations with Stock Returns and Firm Values," *Journal of Accounting and Economics*, December 1997, p. 301–36.

[18] Solomons (1965) research on divisional performance measurement discusses residual income in detail.

[19] Peter Drucker, "The Coming of the New Organization," *Harvard Business Review*, January 1, 1988, p. 1–9.

[20] This section is based on research at the European Telework Development (ETD), Telework Forum –(www.eto.org.uk) and supporting notes on the implications of networked economy distances.

[21] ETD, notes on the implications of networked economy distances (www.eto.org.uk/).

[22] ETD, notes on the implications of networked economy distances (www.eto.org.uk).

CHAPTER 9

Aggregating Forecasts into a Valuation

The Internet bubble was the biggest disruption to real innovation in history.

– Bill Gates[1]

Recessions catch what the auditors miss.

– John Kenneth Galbraith[2]

UNTIDY ACCOUNTING

Bill Gates, arguably our most influential information technology pundit during the heady years of the dot-com frenzy, confirmed our darkest suspicions – that many, if not most, of the departed dot-coms were frivolous, if not outright fraudulent. The Internet bubble sucked up hundreds of thousands of talented knowledge workers, and trillions of dollars of resources, giving both to the wrong people. The bubble inflated on the hype and flawed numbers concocted by a superabundance of contradictory valuation systems.

The Internet did more than just fuel the dot-com bubble in the financial community. The financial community was perhaps the industry most profoundly affected by the growth of the Internet. Whereas up to the mid-1990s, analysts made considerable money selling research reports, the late 1990's saw brokers such as eSchwab and eTrade give better and more extensive analysis away for free. Analysts at large firms were forced to scramble for remuneration from other sources (not always ethical) in order to maintain their jobs. The "celebrity analyst" was born – an essential part of a traveling "dog and pony" show that hyped, hawked, and ramped up the price of the firm's initial public offering.

Auditors were supposed to provide the "quality control" over evangelistic securities dealers, the last line of defense against investor fraud. They failed dismally at properly valuing dot-com and telecoms stocks, despite reams of paper devoted to the complexities and arcanities of accounting. Is there something fundamentally wrong with the way accountants and auditors compute value?

Professor Joel Demski, a noted accounting academic, has observed the tendency of accountants to be more concerned with "orderliness" than with correct valuations – what he calls "tidy bowl accounting." This is the legacy of error control that begins and ends with "double-entry" and is surely at the root of disparagements about beancounting. Professor Demski has a point; and it is one with which most accountants would not disagree. A number of important advances that could have made financial accounting a useful tool for the knowledge economy have been thwarted by "tidy bowl" accounting. Nowhere has this had greater impact than on fixed asset accounting and its associated depreciation. Three approaches have received attention and at least limited acceptance: (1) replacement cost of the asset; (2) its net realizable value; or (3) its economic value to the business. Alternative approaches to accounting for assets come under the rubric of "current cost accounting," "replacement cost accounting," "fair value accounting," and so forth. The most general term, *current cost accounting*, assumes that assets are stated at their value to the business, also known as their deprival value. The deprival value of an asset is the loss that a business would suffer if it were to be deprived of the use of the asset, and could be the replacement cost of the asset, its net realizable value ("fair value"), or its economic value to the business. The most compelling argument for current cost accounting is that a business maintains its operating capacity by separating *holding gains* from *operating income* and thus can tell what part of their success (or failure) to attribute to management or to "environment." The current cost accounting *operating income* is derived by making a number of adjustments in the historical cost accounting *operating income* for the cost of sales, depreciation, and monetary working capital. In addition, the *gearing* adjustment in current cost accounting reduces the charge to the owners for the effect of price changes on depreciation, stock, and working capital. It is justified on the grounds that a proportion of the extra financing is supplied by the loan capital of the business. A *current-cost reserve* account in the balance sheet is used to "collect" the current cost adjustments, and keep accounts "tidy," that is, make sure that debits equal credits. The approach was detailed in the Statement of Standard Accounting Practice (SSAP) 16,[3] *Current Cost Accounting*, which was issued in March 1980 but withdrawn in April 1988 after vehement criticism by the profession. The main objections were that the methods were not "objective," their proxy term for accounting not based on historical costs and arms-length transactions.

More recently, the European Commission has proposed, as a part of its *Action Plan for a Single Financial Market*, to modernize accounting rules by introducing the "fair value" accounting method. The approach is to be limited to derivatives and other marketable, but volatile assets. Nonetheless, a precedent had been set – one that had been vehemently opposed by the profession.

Certainly, wading into this controversy is not for the faint of heart. At the risk of being overly bold or venturesome, allow me to step forward with my own contribution to the fray. I believe that SSAP 16 basically got it right. In the 1980s, accountants could still, perhaps, get away with arguments to the contrary, because the "intangible" assets of technology firms had still not grown near as large as they are today. The majority of wealth created by firms such as Microsoft, Qualcomm, and Oracle today cannot be valued "objectively" that is, based on historical costs and arm's-length transactions. Current cost accounting or some other less "tidy" approach is needed for valuing technology investments.

In *Valuing Technology*, I argued that widely used valuation tools, such as discounted cash flow, while they do a good job of aggregating diverse sorts of financial information, provide no help at all in analyzing the information used to compute values. These approaches are mute on risk, robustness, accuracy, and quality of data or computation. Indeed, many will argue that good valuation is an art, not a science. This is why the advice of professional analysts, fund managers, and other financial professionals is needed – so goes the argument.

The remainder of this chapter addresses the *aggregation* of the value flow forecasts generated by Financial Dynamics into measures of *value* and *return* that are meaningful in the knowledge economy. The first approach computes a *present value* (PV) using a residual income approach to forecast aggregation. The second approach applies Financial Dynamics' to compute a *return on investment* (ROI) for a particular strategy model.

THE SCIENCE OF FORECASTING

Nils Bohr, Nobel laureate in physics, once observed that "prediction is very difficult, especially if it's about the future." The fact is that no one, even if it was within their power, is going to just give us the future value flows of the firm, be they cash flows, revenues, or some other measure of net value inflow. At best, you can only guess at the future. The analyst's choice is whether to make that an *educated* guess by using proper statistical methods. Fortunately, there is a well-developed discipline of econometric forecasting from which we can borrow techniques to make our guessing as educated as possible. Presentation of appropriate econometric methods will consume the majority of this chapter.

Forecasting in disciplines such as physics has necessarily been a part of the "scientific method." Scientific method is the process used to *collectively* construct a *reliable, consistent*, and *non-arbitrary* representation of the real-world. Forecasting is a process of drawing inferences about the future from the present and past. In financial analysis, scientific method implies specific characteristics:

- *Collective:* Financial analyses and reports must be easily understood by others, and their assumptions stated clearly enough to assure that others can replicate their conclusions.
- *Reliable:* Financial analyses and reports can be relied upon to make decisions. Reliability implies not only a *specified degree of accuracy* in financial reporting, but a clear idea of how accurate the reported numbers are through reporting of a *dispersion* statistic – for example, variance or standard deviation – to measure the reliability of the reported value.
- *Consistent:* Financial analyses and reports will not change in the absence of fundamental information. This, perhaps, was the requirement most violated during the dot-com bubble, when analysts fell all over themselves trying to justify ridiculous stock prices. "Consistent" methods would limit and assure that different analysts using similar data would produce similar valuations, or

where they were different, could clearly explain the source of the difference.

- *Non-arbitrary:* The methodology defines a clear role for data, value drivers, assumptions, and hypotheses, which will be applied consistently from analysis to analysis. This latter characteristic recognizes that personal and cultural beliefs influence our perceptions and our interpretations of financial data. We can minimize these biases by using a standard formal methodology which encourages four traits of good research: (1) explains observed phenomena, and explains why; (2) is consistent with previously established knowledge; (3) is verifiable by other parties with access to the same data; and (4) stimulates further discussion, investigation, and revision as data becomes available.

So, what is "scientific" financial research? The simple version in the context of financial reporting and analysis looks something like this:

1. Observe some aspect of the operation of the firm, the industry, or the market.
2. Formulate a model that is consistent with what you have observed in the context of management's policy and strategy choices.
3. Use the model to make predictions.
4. Test those predictions by further observations.
5. Modify the model in the light of your results.
6. Go to step 3.

This leaves out the co-operation between analysts in valuing firms or their projects, and the fact that it is impossible for every analyst to independently have access to every relevant fact or modeling tool. Life is short and analysts have to rely on the work of others. This is why integrity in financial reporting is so crucial – analyses (including the growing number of people who do their own analysis) will often be believed, and most people will not bother to investigate data sources or methods of analysis.

Good forecasts exhibit several characteristics:

1. They provide unique predictions for a given set of assumptions and historical data.
2. Qualitative and quantitative data can be incorporated into the model

via formal model constructs that provide specific roles for both types of data in generating a forecast.
3. They clearly state the degree of accuracy of predictions.
4. They provide the ability to quickly incorporate new information – that is, to "learn" from ongoing events.

A particularly desirable property in forecasting, then, is that the specific forecasts should not change radically too frequently. Otherwise confidence is impaired, communication breaks down, and performance deteriorates. Such models are called "robust" – that is, learning, as new information becomes available, will generally involve routine adjustments and small improvements in the estimates of model quantities.

When significant new information is received, these "exceptions" may involve basic model revision. Respecification of the forecast model will occur when there are specific exceptions. Exceptions can arise in two main ways. The first occurs when "nonroutine" information or events happen which lead to anticipation of a future major change which will not be properly reflected by routine learning. The second occurs when performance monitoring identifies deficiencies in the "routine" model, resulting in questions of model completeness and validity.

The German polymath Gottfried Wilhelm Leibniz[4] repeatedly observed that "*the present is pregnant with the future*". Indeed, the only way that we can predict the future with any accuracy is to extrapolate systems and trends in the present, modulated by our strategic intentions.

Financial Dynamics adopts a philosophy of simplifying wherever possible. Generally, only one parameter is used to describe any particular characteristic or phenomenon. There are two reasons for this. The first is that often these models are being estimated from short histories, with few data points. In most cases, there simply is not enough data to estimate more than the minimum number of parameters. Second, a simple model is inherently more robust, given this lack of lengthy time series, or a large number of data points.

GROWTH

No matter how much we would like, there is no free source of future financial information (unless you believe in time travel or clairvoyance, in which case you may have other books on your reading list). We do our best to *predict* the future by gathering past objective and descriptive information about processes, and consider how trends and inherent uncertainty will project into the future.

There exists a body of research in economics concerned with the characterization of possible growth paths in a competitive economy, or of interesting growth programs resulting from planning.[5] Most of this addresses macroeconomic growth, and the results obtained thus far in the context of firm level and microeconomic formulations have found little application in practice.

Financial Dynamics lends itself well to the incorporation of growth models, if and when they demonstrate their worth for firm-level analyses. But for the duration of this book, I will limit discussion to least squares models with a constant growth rate over the forecast period. Growth is seen as resulting from two sources: (1) growth induced by managerial strategy; and (2) growth induced by supply and product market factors – what we are calling "environment." Because Financial Dynamics forecasts are random variables themselves, we simply assume that events become more and more uncertain as they recede into the future. Discounting or other forms of "ranking" the importance of periods will then be used to determine how to interpret results projected for the more distant future.

OWN INTEREST RATE AND THE OPPORTUNITY COST OF CAPITAL OVER TIME

The transition from a "stored wealth" to a "dynamic value added" model of business challenges accounting's orthodoxy. In the orthodox models, any payment, expenditure, or sacrifice of resources to acquire something is called a "cost." If the economic value of the thing that is acquired expires immediately, this historical cost is called an "expense"; if it expires within a year, the cost is called a "current asset"; if it expires in more than a year, it is called a "fixed asset" or "depreciable asset."

In dynamic models, there is the added complexity of the choice in the forging a particular strategy between early or late recognition of value. Discount rates reflect the price we are willing to pay to exchange value in the future for something in the present. Accountants compute a one-year return on investment (ROI) as *net income after taxes ÷ total assets*. For the longer term, the depreciation and expirations of fixed assets must be taken into account, and a stream of forecasted values of net income after taxes is needed. This can grow quite complex if one is assessing an investment sometime back in history and already knows the resulting income stream; it is generally not possible to use historical financial accounting data to accurately compute the future ROI of an investment today. In knowledge firms such as Microsoft, we are even further hampered by a difficulty in obtaining an accurate accounting for assets. Until recently, all software development was expensed immediately in Microsoft, while income from the sale of that software was accurately tracked. Taken at face value, a computation of Microsoft's ROI would dramatically overstate returns, not at all reflective of reality.

Financial accounting does a relatively accurate job of accounting for current period operating income (although, as Standard & Poor's has shown with their alternative measures, the financial community feels that there is room for improvement). The greatest problem lies with fixed asset accounting and the associated depreciation, which involve predictions of economic life, maintenance, and replacement cost.

The problem with Microsoft and most other knowledge-intensive businesses is that their wealth is no longer retained in an imperishable cache of assets which can be sold or traded when needs arise. Instead, the firm needs to be considered an efficient (or inefficient, as the case may be) machine which generates value through its operations. Value in this context is some function of the stream of future value flows. Discounting is one function which maps future income flows into a present value; it is reasonable to assume that there may be other functions that might actually be more appropriate to specific decision-making situations.

At the core of Financial Dynamics is the *strategy model*, which determines the portion of future value flow that is under management control. Consistent with accounting orthodoxy, Financial Dynamics considers expenditures to be costs. It diverges by not further classifying costs into expenses, current assets, and fixed assets. These artificial divisions had a purpose in the agricultural economy for which they were designed, where production cycles waxed and waned to the ebb and flow of the seasons. They have become arbitrary in the knowledge economy, where firms such as Microsoft may have a hard time figuring out what costs have expired, and which are still yielding new harvests. For example, Visual Basic shows up in Word, Excel, and other Office software as a script editor; it is a central tool of .NET for integrating user requirements;

and it is available as a stand-alone tool for programmer use. What exactly are the returns attributable to the cost of programming Visual Basic?

For firms such as Microsoft, decisions on the software production and marketing consume all of management's time and effort; management of storable assets and property is left to landlords and property agents. In such an environment, concepts of ROI are not particularly useful, because it is hard to identify exactly which assets are generating which returns.

Financial Dynamics uses an implicit opportunity cost of capital for return on investment (ROI). Accounting orthodoxy insists on identifying which costs go with which benefits. This is due to the *Matching Principle*, which is generally applied to assure that expenses and revenues are recognized in the correct time period in order that operating income is fairly stated. Since Financial Dynamics is *strategy-centric*, the corresponding principle would match the expenditures and benefits from a particular strategy item. If we think of ROI as being an annualized interest rate, then we can compute a Financial Dynamics counterpart to ROI as the ratio of: (1) value that will be generated within one year; and (2) value that will be generated after one year.

Discounting involves the choice between present and future consumption: what quantity of a good are you willing to give up today for a promised delivery of one unit of the same good one period in the future? Malinvaud[8] refers to this as one's "own interest rate" for a particular good or service which is typically compared to some "hurdle" rate, or opportunity cost for investing.

Analysts like the idea of an unambiguous interest or discount rate, where the particular good is the *numéraire* – money. In the interest of "objectivity," analysts will try to obtain some sort of market average of individuals' discount rates for money (individuals being people or firms), justifying this as an opportunity cost of alternative investments. Since market rates can be volatile, analysts tend to shy away from choosing one market (say, commercial paper) over another (say, S&P's AAA bonds). Often they would rather default to a market that reflects a predictable set of government policies – for example, Treasury Bonds. Although convenient, the Treasury bond rate is unlikely to parallel one's "own interest rate" for a particular good or service.

Fortunately, Financial Dynamics gives us a convenient alternative to rummaging through external markets for an appropriate internal rate –

one specifically tailored to the experience and stated strategies of the firm. We can define the firm's "own interest rate" under a given set of strategies by the ratio of the value exchanged today to the value received in the next period. There is a convenient interpretation for $\hat{\rho}$, the empirical *strategy lag*, computed by historical estimation of the consequences of management's current set of technology strategies. Empirical *strategy lag* ρ is the correlation of the current residual in our nonlinear least squares regression for θ and the next period's residual. These residuals are determined by management's choice of strategy stated in terms of value (from our *strategy model)* net of the environmental factors θ that we inferred from the regression of the strategy model against historical data. The uncorrelated part of next period's residual is essentially value creation deferred to the next period by management's strategy choice. This, in turn, implies that management's "own interest rate" is approximately $1-\hat{\rho}$; and this compounds, being $(1 - \hat{\rho})^n$ at n periods in the future. This may be considered a return on investment from management's chosen strategy model applied under historical circumstance. Assuming a relatively stable environment (often a big "if" in technology companies), this is a reasonable assessment of future return.

The own discount rate $1-\rho$ is a measure of the one period opportunity cost of capital generated by management's chosen strategy. It is alternatively perceived as an *intrinsic psychological value* – that is, "impatience"; or as an *opportunity cost* – that is, the returns from possible alternative investments. The latter definition is commonly applied, and requires prediction of uncertain future opportunities. Future discount rates are never objective; thus, it may be desirable to compute value based on a schedule of possible discount rates on either side of $1 - \rho$.

THE VALUE CONE

Forecasting (outside the realm of occultists) usually consists of methods for extrapolating current trends to tell us something about the future. Economists, in fact, are fond of the Latin phrase 'ceteris paribus' meaning 'with all things equal' to imply that their forecasts will be accurate as long as current trends do not change. Yet as we have argued, the dynamics of change have undermined traditional accounting in the knowledge economy. These same dynamics threaten our forecasts as well if we fail to accommodate them.

Consider a basic set of 10 historically reported values (the small circles in Fig. 9.1)

Figure 9.1 Components of Forecasting

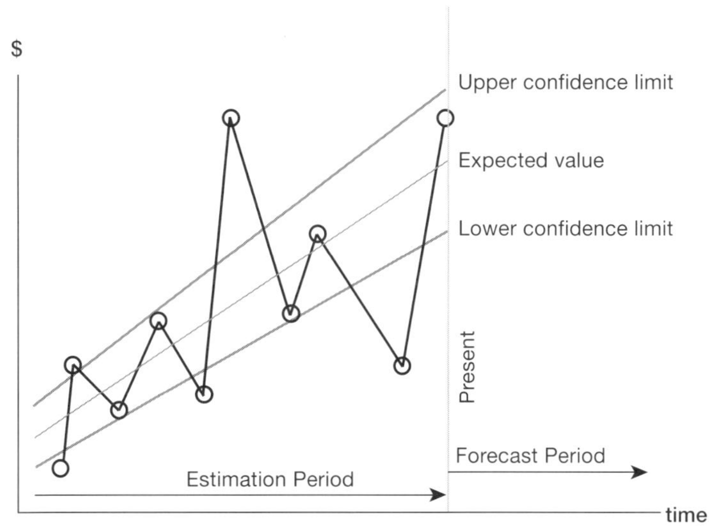

The 'estimation' period over which these values have been reported serves as the time frame for applying the phrase 'ceteris paribus' for the forecast; i.e., we assume that the same things hold in the forecast period as in the estimation period, except where we have specifically incorporated the changes.

Linear regression generates for us a line of expected values over the estimation period, based on the entire set of historical values. From the regression, we may define confidence limits for our estimates reflecting the inherent variability of historical values. Auditors typically choose confidence limits of 70% for account valuations (because that is the minimum confidence limit that their audit sampling literature recommends), but for most purposes, these should be above 90%. Then stripping away all but the outcome of the estimation yields a 'value cone' of possible values. Note that this might be considered one way of smoothing income figures (if this is the chosen metric for value) since if there is an error in one years income number, it will wash out in a subsequent year.

Figure 9.2 The Value Cone

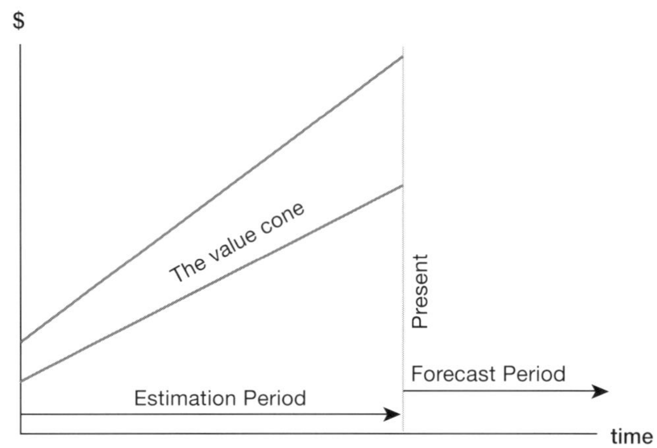

If we assume 'ceteris paribus' without specifying any changes at all in the environment, then the forecasting task is easy: just extend these two lines out to infinity (Fig. 9.3)

Figure 9.3 'Ceteris Paribus'

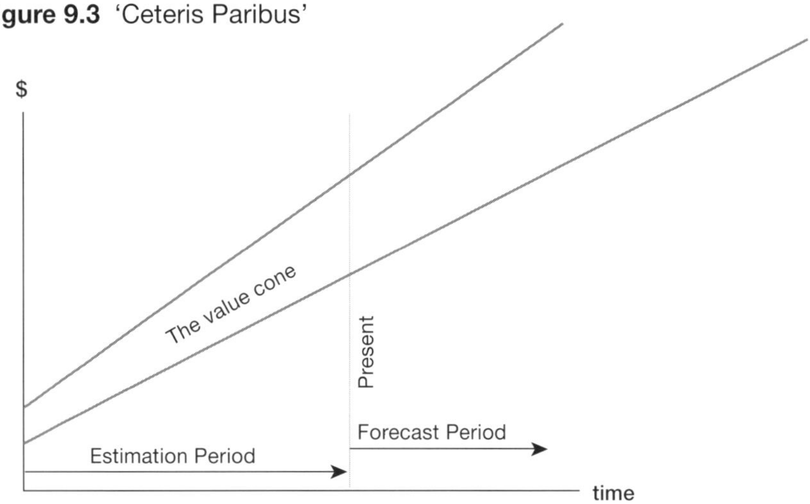

Note that the lines diverge further and further apart, implying that our forecasts become less and less accurate into the future.

In practice, some things change. Changes take place in two areas: (1) *strategy choices,* which influence internal operations, or relationships with outside markets, over which management has some control, and (2) *technology choices*, which are controlled primarily by R&D at competitors labs and universities, by standards, and by markets, all of which are external to the firm. These choices will alter the direction and divergence of the confidence limits defining the value cone (Fig. 9.4)

Figure 9.4 Successful Strategy and Technology Choice

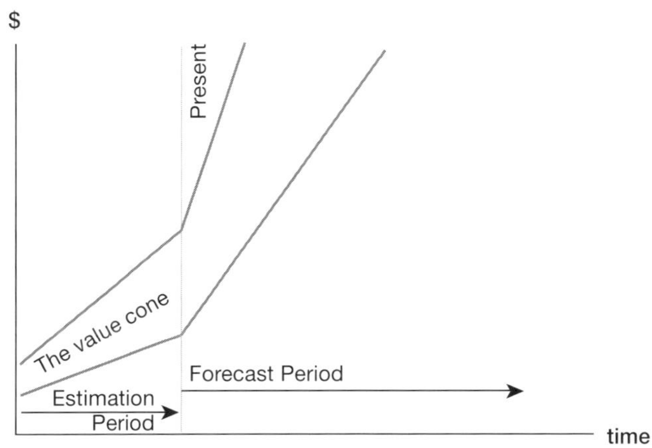

Figure 9.5 Poor Strategy and Technology Choice

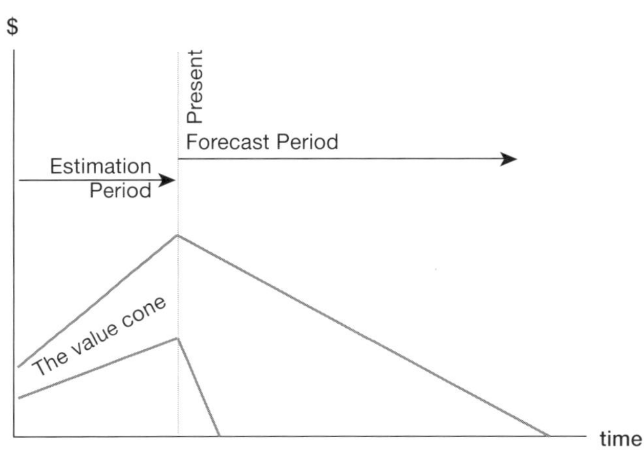

Figure 9.6 High return 'Risky' Strategy and Technology Choice

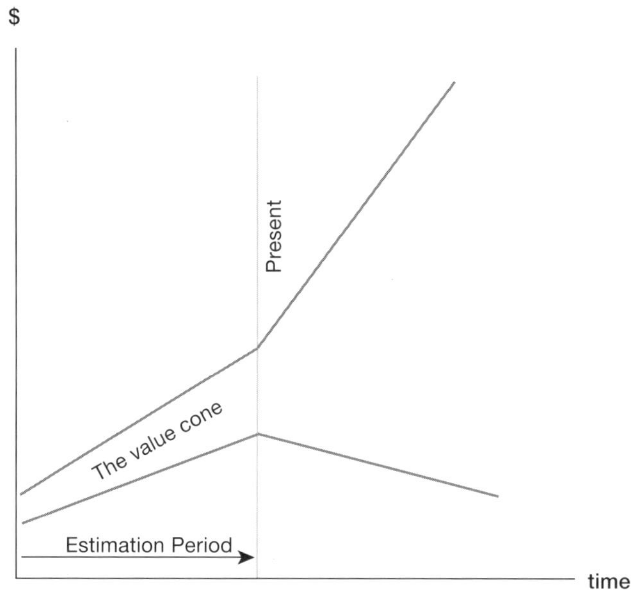

Figure 9.7 Conservative 'Safe' Strategy and Technology Choice

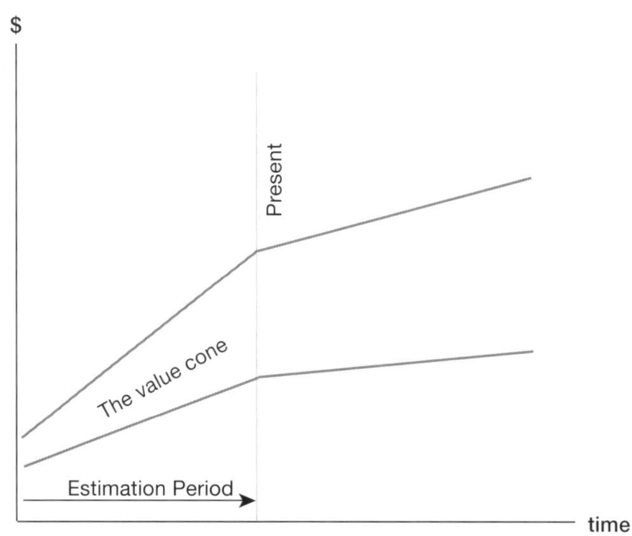

COMPUTING A UNIQUE FIRM VALUE

The value of a firm, project, or investment is the total of all future net value inflows discounted back to the present. Of course, this invites two rather tricky questions: (1) Where can we find "future values"? and (2) Where can we find a discount rate?

A major reason for avoiding amalgams of capital costs such as, weighted average cost of capital (WACC) is to avoid the required proliferation of judgments about the exact cost of a given portfolio of securities, and just how much weight that portfolio should be given. Though bonds and other debt instruments will have a nominal "cost" printed on them, they are also actively traded in markets, and the true yield can vary significantly from that "cost." Management decisions will vary with the yield as well, often leading to refinancing. Take in the additional confusion required by the prediction of future tax rates (also a part of WACC calculations) and WACC provides a recipe for self-deceit.

Consider the example of Michael Milken of Drexel Burnham Lambert. In the early 1980s, Milken worked feverishly to convince traders that risky, high-yield bonds (so-called junk bonds) should be traded (that is, that they actually had some value greater than zero). For two years, from 1985 to 1987 he controlled the market, very likely pushing prices too high (or yields too low) for the commensurate risk. What was the true yield of Milken's "junk" bonds? The topic is controversial, and holds the ability to draw heated remarks nearly 20 years later. Yet things have become even more complex in that period, with an explosion in derivatives – financial contracts that depend in turn on prices or rates of other financial securities, often bonds. Derivatives are the exotic financial tool that cost California's Orange County US$1.5 billion in investments and brought down Barings Bank when Nick Leeson bet against Japanese markets and lost. With these sorts of instruments available, punting on a particular WACC has grown awesomely risky.

In complex ownership situations, alternatives have been developed that discount at a rate that compensates the company for the risk it is bearing. There is, for example, an approach called equity cash flow (ECF), which is simply DCF with a different discount rate. In ECF the discount rate is adjusted for fixed financial claims (for example, interest and principal on debt), as well as for risk of holding a highly leveraged claim to the asset. ECF values tend to be volatile, since capital from many sources is usually pooled because any one of the participants is insufficiently wealthy, or is unwilling to bear the risk of a potential default, or cancellation of the project. Values are likely to change at each interest installment. Not a good thing.

RESIDUAL INCOME

There is a better way. Residual income approaches *only* require the analyst to obtain the "opportunity cost of equity capital," a figure that is needed in the computation of WACC. Equity capital costs are pure in the sense that you are asking how well you could do if there were alternative equity investments available. Better yet, a single equity capital rate is, by definition, immune from the evolving complexities injected by an ever-changing array of available financing alternatives. Note that the effect of any errors in estimation of WACC, where the impact of errors in tax rates and bond yields is complex, can be highly non-linear. In contrast, the impact of errors in cost of equity capital is approximately linear. This can be shown by considering the discounted present value of a stream of value flows $\sum_{i=0}^{\infty} \frac{v}{(1 - r)^i} = v \times r$. Assuming that the stream of value flows does not jump around too much, any errors in specifying the discount rate will have a roughly linear effect on the discounted present value. As the discount rate grows larger, future payments become less important, and only the first few payments account for most of the present value. This actually fits well with the expected decline in accuracy of any prediction of cash flows in the distant future. A model that exhibits this sort of stability of results is called "robust."

The equity capital cost in residual income has traditionally been calculated using the much disparaged capital asset pricing model (CAPM) as follows:

$$r_e = \alpha + \beta(r_m - r_f) + \varepsilon$$

where

r_e (that is, return on equity capital);

r_m is the return on a market portfolio (with diversified risk);

r_f is the "risk-free rate," typically measured as the return on a U.S. Treasury bill with the same maturity as the investment project in question; and

ε is an error term which is assumed to be normally distributed with mean zero.

If appropriate stock-market data are not available – for example, in the case of spin-offs or private companies – β is estimated by looking at comparable public firms. According to the efficient-markets theory, investors who take bigger risks should expect to be rewarded, on average, with higher returns. Since shares are riskier than government bonds (the basis for determining the risk-free rate), it is no surprise to learn that they have generated higher returns for investors than have bonds. One recurrent problem in the use of this formula is that the equity-risk premium has been much larger than the standard CAPM calculation of risk predicts it should be.[9]

Perhaps a larger problem in the use of equity capital cost as the discount rate is that it really ignores the specific investment alternatives that the firm faces. It is an "opportunity" cost, but not necessarily for the specific firm being valued. Rather, it is an average for all the traders in a stock market. Furthermore, there are other "opportunities" than just investment in the stocks of publicly traded companies; and if you are obtaining your numbers from, say, the New York Stock Exchange, then this set of "opportunities" consists only of the top 2,000 publicly traded companies in the U.S. It is a generic measure, but suffers from the fate of many generics – it does not fit the facts of the specific firm.

Fortunately, we have a discount rate for the firm's other investment opportunities in our estimate of $1 - \hat{\rho}$, management's willingness to trade tomorrow's income for today's. Financial Dynamics chooses $1 - \hat{\rho}$ as its discount rate.

OWNERSHIP

Our application of residual income to this point has assumed that we first value the entire business, and then subtract the value of any debt claims and other partners' equity interests. The approach requires managers to presume that they know the true value of those other claims. In practice, they do not know the value of all claims against the firm. To find this value of claims entails an additional valuation methodology.

When a company participates in joint ventures, partnerships, or strategic alliances, or makes large investments using project financing, it shares ownership of the venture with other parties. In such cases, ownership claims can substantially influence the value of the firm, while being contingent on numerous factors. In the extreme, some ventures with negative net present values are sound investments because a partner or lender makes the deal very attractive. For example, governments will deliberately subsidize projects in order to achieve social goals through monopoly concessions, and so forth.

FITTING THE ANALYSIS TO THE DECISION

You can put a value on an entire business by using present value formulas, and indeed this is at the basis of much of our "fundamental" analysis in practice for deriving stock-market prices. But discounting future flows to a net present value provides only a static measure – it compresses the future into a single point in time, losing quite a bit of useful information in the process. Bennett Stewart of Stern Stewart[10] has been a strong proponent of period-to-period comparisons, retaining the information gained from forecasting. Stern Stewart's shift from a single present value figure to period-to-period comparisons has received growing support, and without a doubt provides information needed for some types of investment and management decisions. The analyst should consider whether the decisions which will be supported by Financial Dynamics analysis might better be supported with the undiscounted stream of future value flows. Scenario analysis, in particular, is best supported retaining the time-specific part of the forecasts.

One last rule of thumb: if all your proposed benefits of a given set of strategy choices do not show up until after the third year, then these are probably poor strategy choices. This is even truer if value only shows up after the fifth or tenth or fifteenth year, as was common in late 1990s' analyses supporting many of the now-failed dot-coms. Many things can happen in three years, and if you don't have some benefits kicking in sooner, your assumptions can easily be swamped by forecasting uncertainty. A strategy where the value generation is staggered also gives you some convenient check-in points to evaluate the progress of their implementation.

---

1 Kevin Maney, "America Shows Amazing Resilience," *USA Today*, December 27, 2001, p. B2.

2 Rupert Cornwell, "Shocked and Angry: The Prophet Whose Warnings over Wall Street Were Ignored," *The Independent*, July 1, 2002.

3 The Accounting Standards Committee (ASC) was replaced in 1990 by the Accounting Standards Board (ASB). In its life, the ASC issued 25 Statements of Standard Accounting/n Practice (SSAPs), many of which were adopted by the ASB.

4 Gottfried Wilhelm Leibniz, quoted in Swedenborg's "A Philosopher's Note Book", p. 348, though this quote is more commonly attributed to Voltaire lampooning Leibniz in his play, Candide, where he presents Leibniz as the absurdly optimistic Dr. Pangloss. The full quote is:

> "It is one of the rules of my system of general harmony that the present is pregnant with the future, and that he who sees the whole, sees in that which is, that which is to be. And what is more, I have established by way of demonstration, that in each part of the universe God sees the entire whole, and this because of the perfect connection of things."

5 The seminal work in this area appears in T. Koopmans, T. (1964) "Economic Growth at a Maximal Rate," *The Quarterly Journal of Economics*, August 1964.

6 M. Gordon, M (1962) *The Investment, Financing and Valuation of the Corporation*, (Homewood, IL: Irwin, 1962).

7 H. Pagès, H. (1999) "*A Note on the Gordon Growth Model with Nonstationary Dividend Growth,*" *Bank for International Settlements Working Paper No. 75*, Basel Switzerland, 1999.

8 E. Malinvaud, E. (1985) *Lectures on Microeconomic Theory*, (Amsterdam: North-Holland, 1985), Chapter. 10.

9 The phenomenon was identified in 1985 by economists Rajnish Mehra and Edward Prescott after examining nearly a century of returns for U.S. shares and bonds. Inflation-adjusted return on shares was found to be about 7% on shares, versus about 1% on bonds – a 6% equity-risk premium. In contrast, efficient market theory using CAPM suggested that shares should have traded at only 1% over

bonds – that the extra 5% premium was somehow 'irrational.' Part of the problem is one endemic in valuation. The equity premium is a forward-looking measure, resulting from the extra returns that investors in shares expect to earn. But the equity premium studies estimate expected returns using backward-looking financial data – data that obviously was were not available to investors making investment choice at the time. A recent study by Eugene Fama and Kenneth French has addressed this problem, using a model based on how much investors could reasonably expect from dividends, cash payouts in the form of share buybacks, or liquidating dividends. Using data from 1872 to 1999, they calculate an average equity premium of 3.6% over the risk-free rate for the period. Equities performed particularly well in the half-century from 1950 to 2000, a period that coincides, not surprisingly, with a substantial investment in exchange automation beginning with the New York Stock Exchange's DOT system. Faster execution, transparent and timely corporate information, and widespread investor participation made trading more active, and pushed prices higher.

[10] J. M. Stern, J.M. and J. S. Shiely, (2001) The EVA Challenge, (New York:, Wiley, 2001), p. 18.

CHAPTER 10

Fiscal Archaeology

fis·cal (fĭs-kel) (adjective) Of or relating to finance or finances. (Latin fiscālis, money basket)

ar·che·ol·o·gy (är′kē-ŏl-o-jē) (noun) The systematic inference, study and recovery of material evidence remaining from past human life and culture. (Greek arkhaiologia, or "discourse about ancient things")

fis·cal ar·che·ol·o·gy (fĭs-kel är′kē-ŏl-o-jē) (noun) The systematic inference of corporate financial activity through study and recovery of evidence from the incomplete records provided by an antiquated system of annual financial reporting according to GAAP

In a perfect world (at least from the financial analyst's perspective) we would have access to complete and accurate financial data at the moment they were produced. Instead of getting our data 'piping hot' we usually make due with warmed over TV dinners – long-delayed annual reports, with information aggregated beyond recognition – the bones and pottery stands of fiscal archeology.

Fiscal archeology is of course used facetiously here, a term I have coined to capture the irony of trying to assess the finances of a firm constrained by the limitations of GAAP accounting, historical data, and delayed annual reporting. In a world increasingly looking to the future for insight, accounting gives us last year's numbers (at best) and a choice of audited and dated numbers, or unaudited and slightly less dated numbers (ignoring the possibility that management may go off onto tangents of *pro forma* fantasy). Chapter 11 will discuss the audit process, and specifically looks at whether the delay in reporting caused by the audit is really worth the improvement in audited numbers. I will argue that auditing is in such confusion, and audits such a small portion of the value of the company, that there is little of substance that differentiates audited and unaudited accounting numbers. But the audit *does* assure one thing – that any numbers reported will be archeological relics; artifacts from an earlier period in the history of the company.

As an alternative, it is possible to report numbers when they are captured by the corporate accounting systems, and continually update them. This is the approach that Cisco has taken for some of their more important data. Cisco opened up their corporate accounting for inventory and sales cycles – precisely the accounts on which the greatest part of the audit budget is expended. Such policies put a heavy onus on the accounting and information systems staff. It forces them to audit their own numbers up front (rather than after the fact) under public scrutiny. Cisco has shown that it is possible.

The explosion of corporate information at the end of the 20th century has made *ex post* auditing virtually impossible, and Cisco's approach to audit looks increasingly attractive. In the 1930s it may have been possible to audit transactions a year after they took place, but it is impossible in the 21st century where a massively parallel collection of desktop computers is constantly capturing and churning out the data needed simply to run the corporation. Auditing today needs to be performed at the point and time of data capture, and is best done by interested external parties such as vendors and customers.

Forecasting with the limited information available in published accounting reports is a bit like excavating shards of ancient pottery, and trying to reconstruct a vivid account of the communities and people that cooked with them. The challenge facing the archeologist is to make meaningful sense of the past from what little has survived. Financial analysts today are confronted with a similar challenge: the financial activity reported by firms is often obscure and incomplete as a result of flaws in our antiquated system of financial accounting. Harvard Business School's Robert Merton observes that "accounting is being asked to do something it was never meant to do. It is like a road system built for when cars did 30mph being used by cars doing 60 – and then we wonder why there are accidents."[1]

Financial Dynamics faces a more immediate problem. Today's businesses are unable to produce, on demand, the detailed collections of information that Financial Dynamics needs. Firms have enormous investments in their legacy data collection, summarization, and reporting systems – usually one system serves both management's decision-making and financial reporting. These legacy systems have been dictated by GAAP and double-entry bookkeeping. Today's accounts simply provide a snapshot of a company that gives no indication of the firm's exposures to different sorts of risk in the future. (In contrast, Financial Dynamics can be viewed as a way to articulate *strategy* in a mathematically precise way, and of then running the model of the articulated strategy against prior business experience to determine the implications for future value creation).

In the Middle Ages, people in Europe were intrigued by what they referred to as "magic crocks," pots (cremation urns) that mysteriously emerged from the ground. Flints and polished stone axes similarly turned up as farmers plowed their fields. According to popular belief, these were elf-shot or thunderbolts. The realization slowly dawned in more enlightened minds that these "thunderbolts" and "magic crocks" were in fact the relics from a more ancient time.

Financial analysis is coming out of its own Middle Ages. Financial ratios, present value, and so forth are no longer viewed as undisputed fiscal truths, but as imperfect statistical relics from a much less sophisticated time. From these incomplete records, analysts are required to construct an accurate history of the firm's processes, successes, and failures.

GAAP, however flawed, is the current *de facto system* for corporate valuation. I would not expect this to change in the immediate future, if only for the reason that GAAP and GAAP accounting are mandated by law. But as extensions such as Financial Dynamics prove their worth, we may expect parallel systems to arise for managerial decision-making, just as bond rating, DCF, cost accounting, and managerial accounting arose in earlier eras.

Financial Dynamics requires a lot of new data; data that are not traditionally captured in our financial accounting systems. Probably most difficult is the acquisition of measurements on the strategy drivers in our model. As often as not, strategy is couched in non-financial terms. We have been taught to take a more touchy-feely approach to strategy in business schools, where the subject is often taught in tandem with human resources and organizational behavior.

So, without the systems in place to capture these data, how useful is Financial Dynamics? Very. These data, though not captured by the firm's internal accounting systems, are often tracked by securities analysts who have access to internal documents; and at the industry level by consulting firms. By an educated mix of studied inference, and data collection from multiple sources, it is possible to engage in "fiscal archeology" to construct a Financial Dynamics valuation with existing systems.

USING AVAILABLE AND INCOMPLETE DATA IN FINANCIAL DYNAMICS

Measuring "Value Flow"

Financial Dynamics allows wide latitude in your approach to measuring value – the stuff that flows through the value map. This is necessary if the technique is to serve diverse audiences such as bond analysts, creditors, and stockholders. But to avoid the same pitfalls as corporations have fallen into with *pro forma* reporting, all the particular choices relevant to the value measurement need to be explicitly stated. Unless the analyst has access to very complete internal or proprietary information, it is probably best to stick with value metrics that are commonly captured, audited, and measured in the firm's financial accounting systems. I suggest, then, that *Operating Income Before Tax, Depreciation, and Amortization* is a good default measure to use for value, since it applies directly to operations which are the subject of management's strategy and policies in place.

Another related alternative has recently been provided for leading firms by Standard & Poor's concept of "core earnings," a measure aimed at providing a standardized value metric that they feel more accurately reflects the results of operations.[2] S&P includes many costs that financial accounting often excludes; things like expenses from stock option grants and the restructuring of ongoing businesses, as well as the writing down of the value of certain assets, pension costs, and research and development purchased from another party. It also excludes income not directly coming from operations (and thus management's strategic decisions), such as gains on its pension plan, asset sales, or hedging operations, or such things as income from litigation or insurance. This holds the potential for future improvements in the use of operating income for valuing a company's strategy.

Building the Strategy Model

The *strategy model* is the vehicle by which the *value proposition* of each particular strategy and technology choice is made mathematically explicit. Financial Dynamics does not demand one "correct" strategy model. Rather, particular strategies of interest are collected and regressed against historical performance. The resulting forecasting model will construct forecasts with three components: (1) future value added by the selected strategies; (2) other sources of value or loss, impacted in the "environment" estimator $\hat{\theta}$; and (3) the "strategy lag" estimator, $\hat{\rho}$. This is reflective of any endeavor to better understand a situation. We focus on a few things of importance, while relegating the remainder to the "unknown." Through the "environment" $\hat{\theta}$ estimators,[3] we have a good idea of how much we know or do not know about a business model.

We have already discussed two forms of strategies involving technology choice: (1) *static strategies*, which alter the firm's current competitive position (Chapter 7); and (2) *dynamic strategies*, which alter future competitive position in an evolving and predictable manner (Chapter 8). There is a third form that we have not explicitly discussed, and which is an *ad hoc* category that captures any additional strategies that significantly alter the firm's value proposition. This is provided simply as a catch-all in case there are other strategic choices that do not fit within these categories.

A *strategy model* is stated as an exact *functional relationship* between *value* and the *value drivers* – that is, there is no randomness in the formulation, or said another way, it is non-stochastic. It is the embodiment of management's presumptions about the impact of their strategy choice on value generation. Of course, it will be inaccurate, perhaps a small bit inaccurate, or perhaps completely unjustified. The "goodness-of-fit reality check" is made when the strategy model is regressed against historical outcomes.

Unless there is evidence to the contrary, the strategies are perceived as adding "layers" of value. Thus, the aggregate strategy model just adds the individual strategy models from each specified strategy. This is the

Financial Dynamics counterpart to an approach called *adjusted present value* (APV), also called valuation in parts, *value integration*, and *valuation by components*. It relies on the principle of *value additivity* – that is, robust models will split a project into pieces, value each piece, and then add them back up. The most fundamental split is along the line of *"real" value flows* – revenues, cash operating costs, and capital expenditures; and *"derivative" flows* – that is, those associated with the firm's financing program, values of interest tax shields, subsidized financing, issue costs, and hedges.

Mathematically, the strategy model computes value as the product of two functions:

$P_t = f_P(X_{1,t},\ldots,X_{n,t})$ and $Q_t = f_Q(X_{1,t},\ldots,X_{n,t})$ where

$Value_t = P_t \bullet Q_t = f_P(X_{1,t},\ldots,X_{n,t}) \bullet f_Q(X_{1,t},\ldots,X_{n,t})$. The notation uses P_t and Q_t where previously I had been using the generic Y_t. Because Financial Dynamics separates value and volume effects, in chapter 6 we described the estimation model using the generic Y_t to represent both a single transaction's value $P_t = f_P(X_{1,t},\ldots,X_{n,t})$ and the transaction volume $Q_t = f_Q(X_{1,t},\ldots,X_{n,t})$, implying two separate regressions. This in turn implies that θ and ρ are estimated separately for volume and value. For each of P_t and Q_t we apply the generic formula to compute the "environment" $\hat{\theta}$ and "strategy lag" $\hat{\rho}$ estimators:

$Y_t = f(X_{i,t}, \theta) + u_t$

Where:

$X_{i,t}$'s are *drivers* for the ith strategy and t^{th} time period.

θ is the 'environment,' i.e., other contemporaneous components of value not influenced my managerial strategy. This is an estimated variable.

ρ is 'strategy lag,' the autocorrelation of current value generation with the prior period value creation, where the error terms u_t follow a first-order autoregressive model $u_t = \rho u_{t-1} + \varepsilon_t$ and autocorrelation $|\rho|<1$ and the ε_t's are independent with zero means and independent of the X_t's. This is an estimated variable. The complement $1 - \rho$ is management's 'own interest rate' the implicit rate at which they are discounting the future with their strategy decisions.

A typical model might estimate organizational scaling for a given strategy. The strategy model for organizational scaling might appear as follows for the 'value' component:

$$P_t = 310 \times e^t + \hat{\theta}_p \times 10 \times (\text{labor}_t)^{0.3}\,(\text{tech}_t)^{0.7}$$
$$= 310 \times e^t + \hat{\theta}_p \times 10 \times (X_{1,t})^{0.3}\,(X_{2,t})^{0.7}$$

And the strategy model for organizational scaling for the 'volume' component might appear as:

$$Q_t = \hat{\theta}_q \times 150 \times (\text{labor}_t)^{0.5}\,(\text{tech}_t)^{0.5}$$
$$= \hat{\theta}_q \times 150 \times (X_{1,t})^{0.5}\,(X_{2,t})^{0.5}$$

Where θ_p and θ_q are the estimates for the 'value' and 'quantity' models respectively, $X_{1,t}$ = '*labor*' is the amount spent on labor in period t (which may be past, or may be a future budgeted or planned expenditure) and $X_{2,t}$ = '*tech*' is the amount spent on technology in period t.

Each of these is just one component of the aggregate strategy model which will be a pair of summations $\begin{cases} P_t = \sum_i f_i\,(X_{i,t}) \\ Q_t = \sum_j f_j\,(X_{j,t}) \end{cases}$ because of the assumption of value additivity.

The 20% Rule

One important reason to separate our models for value and for volume is that knowledge-based value is difficult to own and keep. We mentioned earlier that economist William Baumol[4] estimated that an average of less than 20% of the total value from innovations goes to investors in that innovation. The remainder of the benefits spill over to consumers and society at large, in higher quality or utility of products, lower cost, or lowering undesirable side-effects such as pollution. As discussed earlier in this book, an initial approximation for this allocation of value might assume a 20-40-40 split between firm, customer and society. The 20% is the only portion relevant to firm or project valuation.

The Annual Cycle

Financial Dynamics centers on flows rather than stocks of value, and thus in purest form assumes a continuous-time update of data. Legacy accounting systems adopt an annual cycle which cannot easily be translated into continuous-time. Legacy financial systems deliver accounting data to the analyst in annualized chunks – snapshots each year of stocks such as accounts receivable and inventory; annual flux for flows such as income and expense. Fiscal archaeology makes assumptions and estimates that attempt to infer continuous-time flows from GAAP's annual numbers. The results will not be error free, but improve on existing DCF and residual income approaches to valuation.

Startups – companies that have less than three years of operations – confront traditional financial analysis with nearly insurmountable problems. My previous book, *Valuing Technology*, described the sorts of spurious valuations that were common when analysts were not constrained by histories of profits or performance. In contrast, Financial Dynamics offers the analyst specific questions that can be asked of startups (which, by definition, lack a history of operations). This is especially important in the analysis of startup businesses subject to high rates of technology acceleration – like the dot-coms. Startups, no matter how short a time they have been operating, have measurable value inflows and outflows. These will change as the business evolves and matures, and this change, in well managed firms, will be explicitly stated as part of management's strategy and technology choice. Provided a full accounting for value flows and planned strategy and technology choices, the analysis of startups falls easily within the domain of Financial Dynamics, without any of the analytical contortions that proliferated during the dot-com bubble.

CASE STUDY: GENERAL MOTORS

Technology at General Motors

Now (finally) we get down to a practical application of Financial Dynamics, and my example application of Financial Dynamics will assess the future of automotive giant General Motors (GM). You might be forgiven for questioning whether GM is the best example of a "technology company," right? This being GM, you are no doubt expecting to read accounts of square-headed fossils in charcoal suits. Ironically, under septuagenarian Bob Lutz, formerly of Ford, Chrysler, and battery maker

Exide, the fossils are stepping a little sprightlier, and GM is doing some interesting things.

In fairness, the characterization of the automotive industry today as an old-style "heavy industry" is misleading. Material and labor – the traditional stuff of sweating in the factories – account today for less than 10% of automobile cost; around 30% goes for supplier assemblies, and the remaining 60% is accounted for by R&D, software, branding, dealer networks, and other intangibles. That's not all. Cars today carry several times the computing power of a desktop PC, require clever metallurgy, tooling, R&D, and industrial design at the cutting edge of technology. Though its products are tangible, the majority of their cost is not, coming from R&D, dealer networks, and other purely intellectual pursuits.

A brief look at the US economy at the beginning of the 21st century shows an automotive industry that has performed poorly in comparison to the industries that are crucial to the success of its products. Commoditization, inflexible labor, global overcapacity and intense competition have undermined the auto sector's traditional business models. Figure 10.1 shows that the entire manufacturing sector – of which autos are just a part – contributes only about a quarter of the $10 trillion US GDP; and automobiles are increasingly built from technologically sophisticated components. Figure 10.2 shows where US productivity has increased fastest. Ironically, many of Figure 10.2's productivity overachievers provide automotive components – machined metals for frames, bodies and engines, chips for motor and ride control, and of course advertising and insurance are integral parts of the consumer automotive experience. But if any of the benefit of improved productivity does trickle down to the auto industry, it is quickly dissipated in intense competition.

Figure 10.1 Contribution to $10,859 billion US GDP (est. 2003)[5]

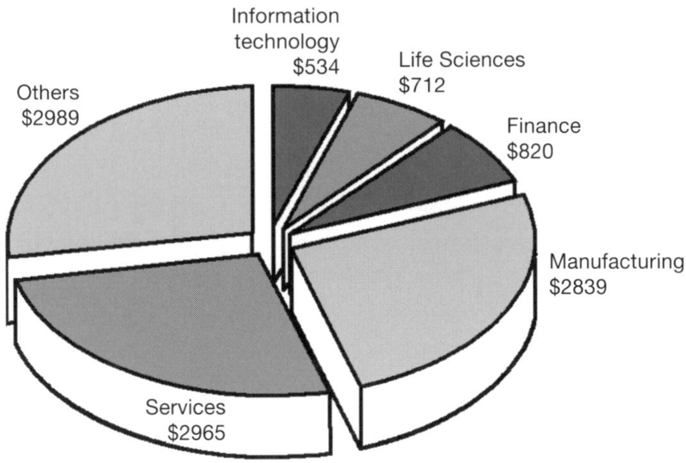

Figure 10.2 Industry Sectors with the highest productivity growth, 1999-2003[6]

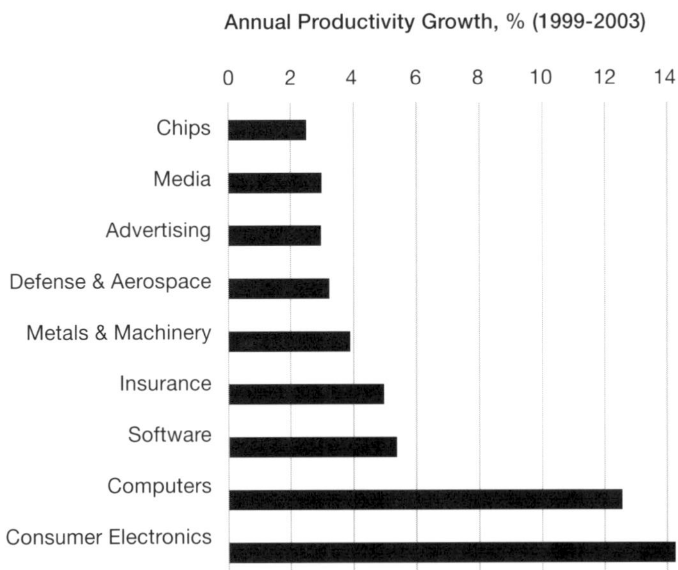

GM itself is one of the most closely tracked firms in the world, and there is a wealth of data on its operations. Furthermore, GM provides their customers with essentially one product – transportation (ignoring the substantial financing arms of these companies, which tend to provide loans for the purchase of transportation). This not only makes it easier to get the data required for Financial Dynamics, but also to "reality check" the resulting valuation against market and institutional valuations.

The automotive industry provides an example of an "old economy" industrial firm whose products have been transformed by technology. They play a huge role in global economics and politics. The world's top three manufacturing companies are also auto companies: General Motors, Ford, and Toyota. Of the top 50 manufacturers, no fewer than 13 are motor companies, employing 2.5 million people. Three times that many are employed in garages and in the industries that supply the car assemblers with parts.

GM, though diminished, remains the world's premier maker of cars and trucks, with brands such as Buick, Cadillac, Chevrolet, GMC, Pontiac, Saab, Saturn, and Oldsmobile (which is being discontinued). GM also produces cars through its Holden, Opel, and Vauxhall units. Non-automotive operations include Hughes Electronics (DIRECTV – which it is selling), Allison Transmission (heavy-duty automatic transmissions), and GM Locomotive (locomotives, diesel engines). GM has a 49% stake in Isuzu Motors and 20% stakes in Fuji Heavy Industries (Subaru), Suzuki Motor, and Fiat Auto (Alfa Romeo, Lancia). It has agreed to take a 42% stake in South Korea's Daewoo Motor. Subsidiary GMAC provides financing. With about 15% of the world's auto market, and about 30% of the U.S. auto market, it does not experience any sort of network externality. Its decisions and strategy choices are driven by general market trends and capacity, as well as by technology and industry practices at all auto firms.[7]

A Short History

General Motors[8] was started in the 19th century out of entrepreneur Willie Durant's buggy business, which he grew into a frenetic exercise in leveraged acquisition and vertical integration. It waxed large and powerful under the leadership of Alfred Sloan (who himself was acquired by Durant along with Sloan's ball bearing factory) to become the world's largest corporation. The company was a pioneer in the top-down command-and-control management style that preceded the influence of widespread digital communication networks. GM's influence mid-20th century was pervasive. At Senate hearings in 1955, Charlie Wilson, the chairman of General Motors, summed up GM's philosophy: "What's good for General Motors is good for the rest of America."

Then came the long period of decline. Ponderous and elephantine under the mantle of command-and-control, GM lost more than half its 60% market share over the next four decades, making ugly and often slipshod vehicles. It missed the rise of the small car in the 1970s, and the sport utility vehicle (SUV) in the 1990s. GM earned its rust-laden reputation.

An impenitent GM is today claiming *not only* to be able to read the post-gasoline future, but to accelerate it as well with new technology embodied in their *AUTOnomy* prototype car.

Can they do it? Car-making is today a knowledge-intensive business. Earlier in this book, I related the story of Yamaha Motorcycle's factory test rider Ichiro Asai. He observed that his contract ensured him a job "until the first time that I crash the motorcycle. Any damage to their prototype and I will be fired immediately." The motor industry spends millions of dollars on R&D, prototyping, and homologation,[9] which can take decades and tens of thousands of test laps. Prototypes are essentially huge, expensive "databases" that are only partly understood at any particular time in their development. Engineers do not like it when a test rider "crashes" their database. But once the design is stable, production is simply a matter of stamping out the finished product at low cost. And the cost of physical materials has been declining rapidly in the automotive sector.[10] Today, the cost of a modern car is mostly "knowledge" gained during R&D, prototyping, and homologation. Expect future technologies to accelerate the tempo of the swing to knowledge economics.

The Decade of Despair

The 1990s, though sometimes profitable, were not good years for GM. GM did not face its "decade of despair" alone. All of the American car-makers have seen their profit margins erode considerably, largely as a result of the Japanese onslaught which started in the 1980s, and has resulted in numerous U.S. production facilities, and one manufacturer,

Toyota, whose main market is now the U.S. Japanese manufacturers have traditionally had profit margins around 5%; while European car-makers have seen their margins vary between 0% and 15% from 1980 to 2000 (Fig. 10.3).[11]

Figure 10.3 Pre-tax return on net assets, U.S. automobile sector;

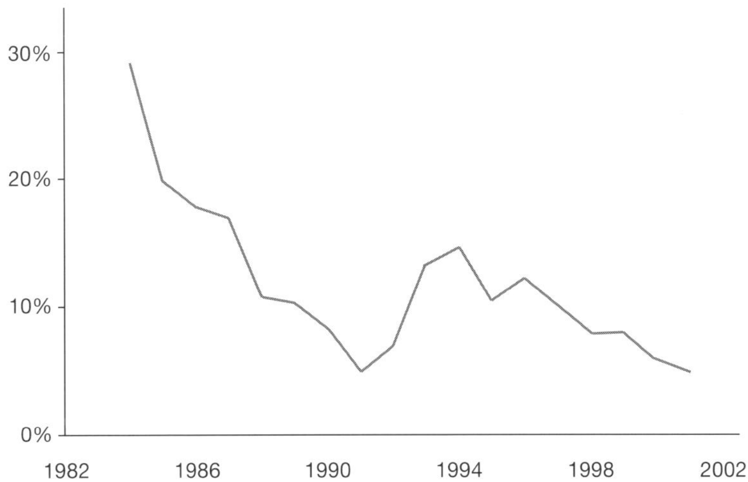

Source: Economist Intelligence Unit.
(Note: GM's ROI on net assets is currently less than 1%)

The trend continues. New car consumer prices declined 1.8% from May 2001 following a 1.7% year-to-year decline in April and a 1.5% drop in March 2002. Each 1% decline in the CPI costs GM around US$120 million, Ford US$100 million, and DaimlerChrysler US$60 million. If all car firms in the world in 2002 ran flat out, they could produce 68 million cars a year (including other light vehicles such as pick-ups and SUVs) The industry currently runs at only about 75% of capacity. Even if every factory in North America could close there would still be excess capacity.

Up until 1990, whenever U.S. capacity utilization exceeded 70%, the return on sales of the big three car firms was at least 8%. But that relationship broke down in the 1990s as cars and supply chains became technologically more sophisticated, and the net profit of GM and Ford fell below 5%. The manufacturers lowered prices, added extra features at no extra cost, and flooded the market with cheap, formerly leased cars.

The cost structure at GM changed along with the rest of the automotive industry; as technology progressed, customers responded to greater choice, which meant more competition for auto-producers. From 1990 to 2000, value generated by each car sales transaction was squeezed as the average sales revenue from a new U.S. car increased over US$4,000, while the firm's cost of providing that car increased by over US$5,000. Of the US$5,000 additional cost, only US$1,500 was for additional parts and features in the car, another US$1,000 resulting for higher supplier prices for parts, and US$2,000 to meet government regulation (what the industry calls homologation).[12] Unfortunately for GM, during the same time period, their own market share fell from 36% to 28% of the US car and light-truck market.[13] For the decade, the statistics were: $\Delta P_{1990\text{-}2000} = -\$1000$ and $\Delta Q_{1990\text{-}2000} = -8\% \cdot market$ (where Δ is the standard mathematical representation for *change* in value over the period).

The Threat of Substitutes

News after GM's "decade of despair" didn't improve in January 2001. That is when a blizzard of rumors hit the press. Amazon CEO Jeff Bezos said it's a "product so revolutionary, you'll have no problem selling it." Apple Computer CEO Steve Jobs said "it will change the ways cities are designed." Harvard Business School Press paid US$250,000 for a book by Steve Kemper about the mysterious invention with the code name "Ginger," yet neither the agent nor the publisher knew what "Ginger" was. Venture capitalist John Doerr claimed that "Ginger" would be as significant as the development of the World Wide Web, and would be the most lucrative start-up in history, making Dean Kamen, its inventor, richer than Bill Gates.

The buzz was about a new challenge to the reign of the automobile. It was the Segway personal scooter with "dynamic stabilization" – a fancy system that ensured that it didn't fall over. The Segway was forecasted to cure most pollution and congestion caused by cars, extend the range of employee activity in closed spaces such as factories and offices, and provide a variety of other transportation benefits.

GM and its rival auto firms looked on with bemused curiosity as the extravagant claims multiplied. The automobile had been challenged before.

The first of the 65-pound US$9,000 Segway scooters were delivered in early 2001. They were truly impressive, with some high-performance units even being combat-tested by the military, presumably off-road and at high speeds. Then the inevitable happened, leaving personal injury lawyers to salivate at the prospects. The heralded Segway claimed its first victim when a member of the Central Atlanta Progress Ambassador Force toppled from his personal scooter at 8:40 pm, Thursday, April 3, 2001.[14] The accident awakened the public to the reality of the Segway. Like a motorbike, it leaves the driver wide open to bodily injury when things go wrong.

The Segway joined the ranks of dozens of other pretenders to the standard for personal transportation – motorcycles, planes, helicopters, hovercraft, bicycles, and the like – as the automobile continued its unbroken reign as "king." Others might claim market niches, but nothing else has the prevalence to remain the standard bearer.

There are more obstacles to entering this market than are commonly supposed. The automobile, at the dawn of the 21st century, benefits from a century of government spending and regulation that ensures it a competitive advantage in the transportation marketplace. This is important, because by itself, the automobile is not the most efficient mode of urban transport – mechanical or biological – as Figure 10.4 shows. The fact is that bicycles provide the most energy-efficient form of transportation: 350 calories (a bowl of cereal) powers a person enough to travel 10 miles; the average auto uses about 18,000 calories for the same trip. Petroleum supplies about 97% of the energy for transportation, making the developed economies dangerously dependent on politically and socially unstable oil producers.

The bicycle creates no air pollution and takes up little space. What's more, the average auto trip in the United States takes one person approximately five miles; distances that could be covered on bicycles if the infrastructure for easy bicycle travel were available.

Figure 10.4 Energy efficiency of transportation

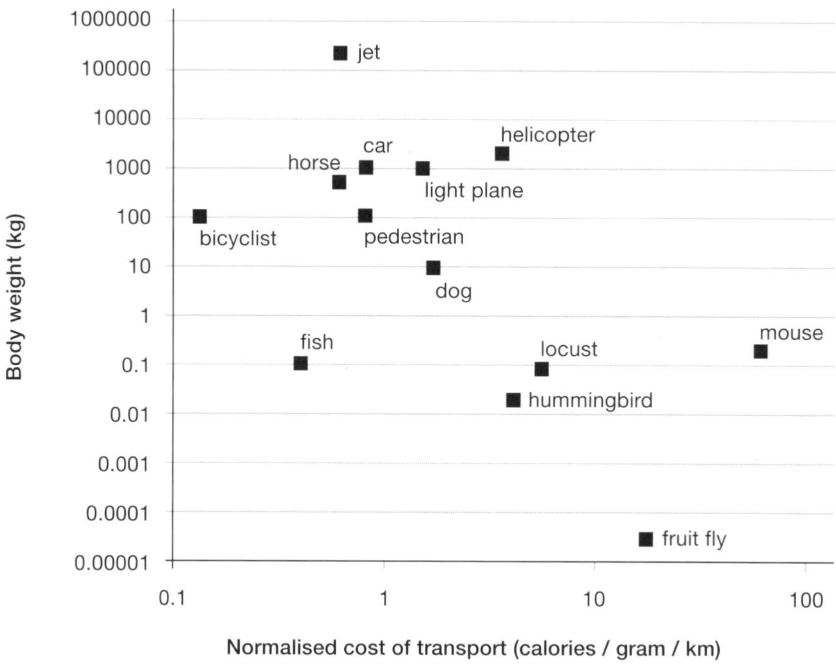

Cars, in contrast, are large, heavy, fast-moving vehicles that demand enormous amounts of space to function effectively. They also have a zone of influence that expands as speed and the quantity of traffic increases, a fact that dictates sprawling land development. Even if cars grow more efficient, they dictate an inefficient use of land and other natural resources, time through traffic congestion, and contribute to social fragmentation.[15]

The automobile would be an unlikely transportation competitor without extensive government subsidies, which will no doubt extend into the foreseeable future. U.S. government subsidies for highways and parking alone amount to between 8% and 10% of U.S. gross national product, the equivalent of a fuel tax of approximately US$3.50 per gallon. If this tax were to account for "soft" costs such as pollution cleanup and emergency medical treatment, it would be as high as US$9 per gallon. The cost of these subsidies – approximately US$5,000 per car per year – is passed on in the form of increased prices for products or, more often, as income, property, and sales taxes. About half of U.S. air pollution emissions come from motor vehicles.[16]

These subsidies, substantial as they might be, probably do not directly enter into strategy considerations, simply because they are so pervasive and widely accepted that they are unlikely to change in the near future. They present substantial barriers to entry that assure that any threats of new entrants, such as the Segway, are kept firmly at bay.

Automotive Technology Acceleration

In 1990, it took car companies five years to develop and start making a new model; a decade later, most could do it in less than three. The car companies are realizing that they have to wheel in fresh products and retire old ones after only a few years. In order to satisfy consumer tastes and fads, most manufacturers are trying to shorten the order-to-delivery cycle time to under two weeks, so that a consumer can get the specification and color he or she wants quickly, rather than have to compromise with whatever happens to be in stock. The trick is to manage the flow of such products so that offerings in the marketplace always remain reasonably fresh, without wasting profits on excessive investment in updates or redesign. If a car company gets this aspect of the product cycle wrong, it soon suffers. In 2001, the auto industry average product age is about four years. Following Volkswagen's lead, industry may be moving beyond the conventional mass production upon which its fortunes are based. Since the 1920s, the way to make a car has been to construct a metal shell bearing all the stresses, based on load-bearing floors; pillars and panels hammered out by huge presses costing millions of dollars.

At volumes of over 250,000, this will probably remain the best approach. For smaller runs, however, a new technique is coming in. With space-frames, car-makers can clip extruded tubes or sections made of steel or aluminum on to a basic floor pan. Cheaper, lighter panels, increasingly made of plastic, can then be hung on this skeleton. With this approach, smaller production runs of models can become viable. Innovations in automotive technology are increasing at an accelerating rate, and this has a significant effect on the future value generation of the product.[17]

Portable internal combustion power trains essential to the automobile took a long time to evolve. The Dutch polymath Christian Huygens originally designed and built a gunpowder-powered internal combustion engine in the 1680s.[18] His research assistant Denis Papin, who was responsible for actually operating these machines, prudently inferred that he might lead a longer and healthier life if, instead, steam were used in a bulky "external combustion" arrangement (both water and gunpowder expand by the same amount, 1,300:1). Thus, internal combustion technology languished until the late 19th century. The U.S. staged its first motor race, in which all vehicles were called "motorcycles" regardless of the number of wheels, from Chicago to Waukegan, Illinois, in 1895. Everything *but* gasoline was used as a power source – acetylene, compressed air, city gas, electro-turbine, ether, kerosene, naphtha, and carbonic acid, to name a few. When gasoline automobiles finally came about, they had hot plug ignitions – a platinum tube that was heated with a blowtorch before the car could be started. Spark plugs, points, and coils quickly replaced the unreliable hot plug, improving efficiency. Early cars had neither gearboxes, nor clutch, nor brakes, but rather depended on engine pressure for what little control was had over power delivery (making them especially dangerous, too). Engines could only be safely revved to 1,000 rpm; Cadillac and Buick quadrupled horsepower by building engines that could rev to an unheard-of 4,000 rpm.

Past industrial revolutions have seen explosions of technology that drive the prices of key commodities (trains, rails, the Model T) down by 5 to 10% per year for several decades. Future technologies will accelerate even faster. For example, Yale's William Nordhaus estimates that the price of computation has dropped 42% per year over 60 years – a trillion-fold fall since 1940.

Using automobile purchase statistics provided on the website of the US Department of Commerce's Bureau of Economic Analysis, and a review of the history of the automobile to compile a timeline of average horsepower ratings claimed by manufacturers, gives the following price-performance relationship

$$In\ P_t = t \cdot (\$0.03/yr - 2hp/yr) + 2.3 \text{ implying that } P_t = 10 \cdot e^{0.14t}$$

where P_t is measured in brake horsepower (BHP) per 1966 constant-dollars, and t is in years since 1979. By our prior convention, the coefficient of technology acceleration for automotive performance in horsepower is $\alpha = -0.14$ horsepower per dollar per year. We will presume that this coefficient will remain stable into the future. To convert this to a 'value' formula, it is necessary to project a power need for vehicles – 200 BHP seems a reasonable figure for GM's fleet – and project the cost of providing this per car. Target BHP not only determines specifications for the engine, but the rest of the power train and chassis. The 1979 average car price was approximately \$6000 and delivered about 100 BHP; by 2000 this was \$20,000 in current dollars, or slightly less that \$6000 in 1979 dollars, at about 200 BHP. Then assuming the target of 200 BHP into the future (though indeed this may change as well), we can predict a decline in average unit cost to produce a car of $\$600 \cdot (1 - e^{-0.14(year-1979)})$ assuming that we hold to 2000 standards of power, instrumentation quality and so forth. That is a big 'if' and further consideration is warranted as we try to visualize the automotive future.

The 'hedonics' in the automotive market are essential to proper forecasting of technology acceleration as it will play out in practice. Instrumentation, fuel economy, power and many other qualitative factors have changed dramatically between 1979 and 2000. These are the qualitative improvements that do not help the profitability of GM – rather they are a cost of staying competitive – and are passed straight through to the customers. We handle these qualitative changes collectively with the '*20% Rule*.'

GM provides us a glimpse of their vision of future automotive technology. Just as power trains have evolved over the life of the car, providing the impetus for continued performance-price improvements since the automobile's early days, so will they continue into the future. GM has invested billions in their prototype for the next anticipated technology leap – a fuel-cell hydrogen-powered car called AUTOnomy. AUTOnomy dispenses with the engine, transmission, steering wheel, and gas tank. With few moving parts, it can last for decades; it can generate more power than it uses, and can be equipped to apply the surplus to power the owner's house. Manufacturing costs a fraction of what it takes to build a traditional car, since the car itself is just a flat, five-inch-thick fuel cell with four electric motors attached at the corners and computer-controlled, drive-by-wire steering. Individual models can be customized and bolted to the fuel cell/floor-panel. More powerful cars will stack fuel cells into a thicker floorboard. New technology promises new problems to accompany rewards, so the move ahead will not necessarily be smooth. The Boston Consulting Group forecast that fuel cells will power one car in five being produced in 2020 (only four product cycles away for a car company); many expect electronic steering and braking to come much earlier. Mercedes thinks it will take about 10 years before the young car buyers of tomorrow, brought up on video games, will accept electronic steering. Given this history, we may expect this technology in the future to accelerate at the rate of α = -0.14 horsepower per dollar per year.

GM still has a way to go. An internal combustion engine is cheap to produce, costing only $50 per kilowatt; most fuel-cell companies project prices of $3,500 to $5,000 per kilowatt, whereas GM is targeting a price of $350–500 a kilowatt. On the other hand, operating costs of internal combustion are volatile, based on pollution and greenhouse gases. Recent Californian legislation ups the ante for internal combustion.

Organizational Scaling

Automation of car-making began in earnest during the second decade of automotive production. The Model T was the first automobile whose production was systematized and automated. In 1915, it sold at half of the original price in 1908, and was a better car as well. There was no obvious increase in material, labor, or overhead invested in each car. (In fact, the latter two obviously dropped.) Instead, Ford's systemization and automation of production followed the principles of Fred Taylor's Scientific Management.[19]

Since the 1980s, the automation of assembly lines has continued with emphasis on data communications. In 1980, GM began development of the Manufacturing Automation Protocol (MAP) specifically developed for computer communications in a factory environment over a Wide Area Network to interface with manufacturing robots, factories for just-in-time parts delivery, and for made-to-order car purchases for dealerships. The MAP protocol, developed in Dearborn, Michigan, influenced the architecture of the Internet through the MERIT network in nearby Ann Arbor, Michigan, which at the time was managing much of the Internet infrastructure. Unfortunately, by 1985, only 15% of GM's 40,000 programmable devices could communicate with MAP, and the standard languished for factory automation, though its influence on the Internet lives on. Subsequent architectures have seen greater success, and now generally use the Internet as a platform.

Table 10.1 Estimation Window Data for Organizational Scaling Model, 1988–2001

Year (t)	Volume (Q) ('000)	Capital (K) (depreciation and amortization) ($millions)	Factory labor (L) net of "non-value added"[1] ($millions)	Automotive deflator (1980 dollars)
1988	8,265	7,338	48,605	115
1989	8,019	7,431	49,983	118
1990	7,901	7,653	51,554	122
1991	7,310	8,110	53,716	127
1992	7,373	9,077	47,864	129
1993	7,851	9,533	53,241	132
1994	8,379	10,323	64,108	137
1995	8,324	12,099	57,258	140
1996	8,381	11,840	66,547	141
1997	8,565	16,616	70,437	142
1998	8,149	12,201	69,522	138
1999	8,786	12,318	71,472	138
2000	8,597	13,411	72,465	140
2001	8,560	12,908	64,044	140

Technology acceleration will affect the rate at which labor and capital are traded off over time, though to my knowledge there is no empirical evidences which could support the definition of a model incorporating time. So for the purposes of this analysis, I will constrain my articulation of GM's organizational scaling model to trade-offs in capital and labor that take place as a result of future planned expenditures (as reflected in the annualized expenditures entered in table 10.1). Once the estimation window data has been collected, we want to regress the data in table 10.1.

In this analysis, assume that labor unions and pensions make it difficult to change the rate of machine-labor tradeoff over the forecasting horizon we use for the GM analysis. A Cobb-Douglas production function would suggest the impact of automation over time is minimal, and that $a \gg b$ (based on table 10.1). This is another way of saying that technology acceleration does not have an opportunity to impact General Motors value creation because of organizational constraints. The main driver of organizational scaling – the improvement in production efficiency because of technology acceleration – is not allowed to benefit General Motors because of the resistance of organized labor. Thus the output impact of tradeoff between labor expense and automation is

$$Q \propto K^{a+bt} L^{1-a-bt}$$

Rearranging, we have

$$lnQ_t - lnL_t \propto (a+bt) \cdot (lnK_t - lnL_t)$$

Use the automotive deflator to convert L and K to 1988 constant dollars and regress to get GM's Organizational Scaling Cobb-Douglas function $Q = K^{5.3} \times L^{-4.3}$

Total spending on labor and capital can be predicted from the forecast formula which solves the budget problem $L + K \leq Budget$ where the annual budget represents a discretionary reduction in value each year.

Geographic Scaling

Three factors potentially affect the way in which technology will squeeze value out of automotive production in the 21st century: (1) continued globalization; (2) improved supply-chain automation; and (3) improved customer relationship management. We will consider each in turn.

■ Globalization

By turning themselves into global firms, car-makers have cut costs and improved efficiency. Just about every volume car producer is rationalizing the number of basic car chassis (known as platforms) it produces, and making each one capable of supporting different models. The same is happening to engines and gearboxes. Thousands of suppliers are being cut to a few hundred, who are being asked to design more new products. Frequently, they must now design and supply, say, whole braking systems, rather than just drums or pads to be bolted together in the factory where cars are assembled. At the same time, concentration in the industry is limiting choice, as fewer and fewer big competitors produce components.

Auto-makers are differentiating their products at the customer interface through a wider range of products and options, custom building of new products, and other approaches made popular in the personal computer market.

■ Supply-chain Technology

A program called the *3DayCar* initiative, sponsored by the major auto producers and run by Cardiff and Bath universities in the UK, found that of the average 56 days that an automobile spends in production, less than five days are actually spent working on any single unit. The rest of the time, production is waiting for information to catch up. The industry feels that, in theory, a car could be produced in three days, given optimal supply-chain management. Today, it takes 60–100 days to make a car, from stamping out metal to delivering the vehicle to the customer. With such long lead-times, car companies push their vehicles out into the distribution network, based on guesses from dealers about what they will sell. A certain percentage of guesses are wrong, resulting in reduced sales revenue. Even if they were all right, the enormous amount of inventory in the supply chain would be costly to finance, potentially wiping out existing profit margins (which we earlier saw tends to hover around 5% of sales).

Manufacturers such as GM and Toyota are planning over the next two decades to phase in a series of production lines that provide a "five-day car," essentially one built to order. So, assume that the supply chain drops from 55 days today to five days in 20 years, implying a 2.5-day per year (4.5% per year) drop in inventory in the supply chain. Assume that GM is basically making the right marketing decisions, or that where they aren't, they have alternative outlets such as rental fleets like Avis. GM's inventory in 2001 was around \$10 billion; so figure \$450 million per year reduction in inventory. Assume cost of capital is 20%, then figure around \$90 million additional operating income generated by each 2.5 day drop in the supply-chain.

For 2001's production of $Q_{2001} = 8{,}083{,}120$ vehicles, an average effect on individual transaction price (each transaction is the sale of one vehicle) of

$$value = (year - 2001) \cdot \$11.13 \cdot Q_t$$

■ Customer Relationship Technology[20]

The Internet allows car-makers and their suppliers to share information in real time, with GM and the other Detroit auto-makers partnering in the Covisint portal. This, in turn, enables them to meet customer demands faster, just when the car industry is considering other programs for becoming more responsive by, for instance, building cars to order. Automobiles are most people's second-biggest purchase. If auto-makers can develop relations with consumers via the Internet, they can respond more directly to demand. Whereas only 3–5% of cars sales are made online, nearly 50% of car sales are influenced by Internet-based consumer research. Firms such as Autobytel and CarsDirect.com in America are already selling cars on their websites in collaboration with dealer chains, which will actually drive the consumer's purchase to their home.

The impact on sales should in general make selling more competitive, giving consumers more choice and service. This will be more expensive for the auto industry, but will be the necessary ante to play in the market. We will make the assumption that GM, being a leader in the market, will choose to play, and that the net effect will be a "wash." Enough revenue will be brought in to cover costs, but the main benefits will accrue to consumers, not auto-makers.

FORECASTING

Now let's aggregate the prior analyses into strategy models. All of the *static* strategic factors were considered not to influence future value in a managed way, and thus will be excluded from the analysis. This leaves us with three *dynamic* strategy factors – geographic scaling, organizational scaling and technology acceleration – which will be added to construct the strategy model. The value effect of geographic scaling is $value = (year - 2001) \cdot \$11.13 \cdot Q_t$. The value effect of technological scaling is $value = \$600 \cdot (1 - e^{-0.14(year-1979)})$. The effect of organizational scaling is split between the value and the volume models. The model we investigate in this chapter has a simple value map (Fig. 10.5)

Figure 10.5 GM's Value Map

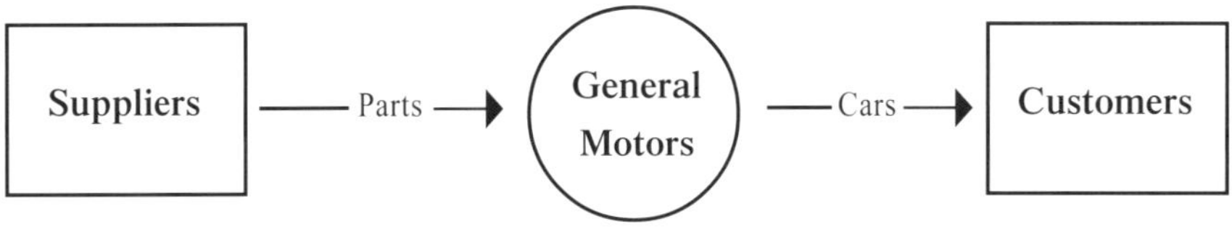

Each 'Cars' arrow generates value that is subject to 'supply' and 'demand' trade-offs induced by constraints on the incoming and outgoing arrows. Typically these would be plotted with supply and demand curves plotted on axes of value *P* versus volume *Q* (Fig. 10.6)

Figure 10.6 Choosing a strategy involves making choices about value and volume

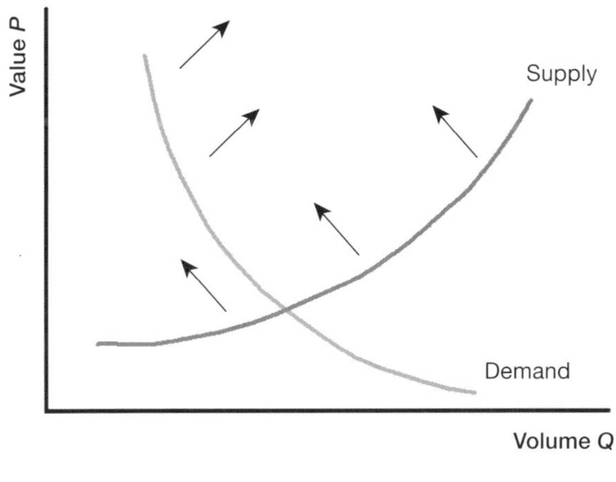

Setting up the strategy model involves making choices, which is more or less what management is all about. And really the first choice in every strategy component involves identifying where the firm intends to operate on either the supply or demand curve (depending on the nature of the strategy component chosen). Although there are theoretically infinite shades of variation, these can be simplified at the extreme into strategies which target a 'commodity' (high volume low unit value) market versus 'luxury' market strategies (exclusive, low volume, with high unit value).

I will make the somewhat subjective assumption that GM has targeted the 'commodity' end of the market, given its poor showings at Cadillac, Opel and Fiat when compared with their rivals at DaimlerChrysler, BMW and even Ford with its stable of Volvo, Jaguar, Land Rover, Aston Martin; not to mention the Japanese with luxury marques like Lexus and Acura.

Note that there is no 'one' right formula for the value map or the strategy model. These basically set out the domain and range of activities that the analyst chooses to collect data on and considers important. In fact, where the value of one strategy over another is compared, the model may be very abbreviated indeed. It's a matter of focus; the center of the model (value map or strategy model) is whatever the analyst chooses to analyze. Financial Dynamics is robust; whatever activities or relationships are excluded from these parts of the model are picked up the 'environment' and 'lag' estimates. The model makes the best of whatever information it has. This is just one of many alternative strategy models that could be specified. This is a very simple strategy model in that it includes no value drivers besides '*year*.' It could be more or less involved than stated here, but this brings together the specific decisions and assumptions we have made about what is important for GM in the future. We state the strategy model in a non-stochastic (i.e., non-random) fashion, because it is simply a working decision that we make in the process of deriving a valuation. Randomness, real data, and forecasting are all introduced in the next step. The strategy model should be perceived as a set of assumptions that management makes about the control they have over future activities. Given a dataset of prior outcomes, Financial Dynamics gives the tools to analyze what the impact of these working decisions are expected to be.

Assume that both the financial and non-financial operations are tied to the automotive industry sales. Table 1 provides a historical time series of value flows and transaction volumes for GM. Value flow is measured as operating income before depreciation, amortization and taxes; volume is the number of unit vehicle sales. This fits with the strategy model we created previously.

Table 10.2 GM's Annual Value Flow and Volume Figures[22]

Year	Operating Income ($millions)	Quantity Sold
1989	20444	8019000
1990	15862	7901000
1991	12257	7310000
1992	13173	7373000
1993	17060	7851000
1994	21565	8379000
1995	25065	8320000
1996	20169	8381000
1997	23978	8597000
1998	21568	8149000
1999	26049	8786000
2000	25373	8083120
2001	25533	8083120

The summary statistics (based on table 10.2) for the estimation period are:

Total Value (Operating Income): μ = \$20,622,800, σ = \$4,743,180,
$\sigma \div \mu = 0.23$

Volume (Quantity of Vehicles Sold): μ = 8,094,790, σ = 426,831,
$\sigma \div \mu = 0.05$

Value (Income per vehicle sold): μ = \$2,530, σ = \$497,
$\sigma \div \mu = 0.2$

This is not really the value per transaction, but the total operating income in millions, which must be divided by the volume.

In addressing GM's strategy choice, GM's Value Map gives us only one set of value equations to compute, those for the 'Cars' arrow. This set contains two strategy models (assume that any costs from 'Parts' arrow are already imacted in the 'Cars' arrow values')

$$P_t = \text{'}\textit{strategy model for value}\text{'} + \theta_{P_t} + u_{P_t}$$
$$Q_t = \text{'}\textit{strategy model for volume}\text{'} + \theta_{Q_t} + u_{Q_t}$$

Strategy models are non-stochastic; they are treated as if they are a statement made with certainty. In fact, they are the equivalent of a working hypothesis, and the Financial Dynamics strategy equations are used to explore the value implications of that hypothesis in the context of the data (i.e., historical volume, value and strategy drivers, as well as planned values of strategy drivers).

Let's make the following decisions with respect to each of the strategy components (all of which, in this example, are dynamic). Geographic scaling will influence value, but not quantity because it allows global choice of both markets and sources; technology acceleration will similarly influence value as GM's management chooses to direct improvements in technology towards regaining market share through lower costs and prices; and organizational scaling influences both value, through the spending on L and K, and volume, because L and K determine the capacity for output, which GM intends to use to its maximum (otherwise why invest in it). Restating:

$$P_t = \text{'Geographic scaling'} + \text{'Organizational scaling'} + \text{'Technology Acceleration'} + \hat{\theta}_{P_t} + u_{P_t}(\hat{\rho}_P) =$$

$$= \frac{(\text{year-2001})\bullet\$90{,}000{,}000}{8{,}083{,}120\ (=Q_{2001})} + \$600\bullet(1 - e^{-0.14(\text{year-1979})}) + \hat{\theta}_P + u_{P_t}(\hat{\rho}_P)$$

where the carets (^) over the values indicate that that value was computed from the data provided in the historical (estimation) period; and K_{year} and L_{year} are the capital outlay and labor expenditure for a particular year,

$$Q_t = \text{'Organizational scaling'} + \hat{\theta}_Q + u_Q(\hat{\rho}_Q) =$$

$$= K_{year}^{5.3} \times L_{year}^{-4.3} + \hat{\theta}_Q + u_Q(\hat{\rho}_Q)$$

Figure 10.7 General Motors Value Cone

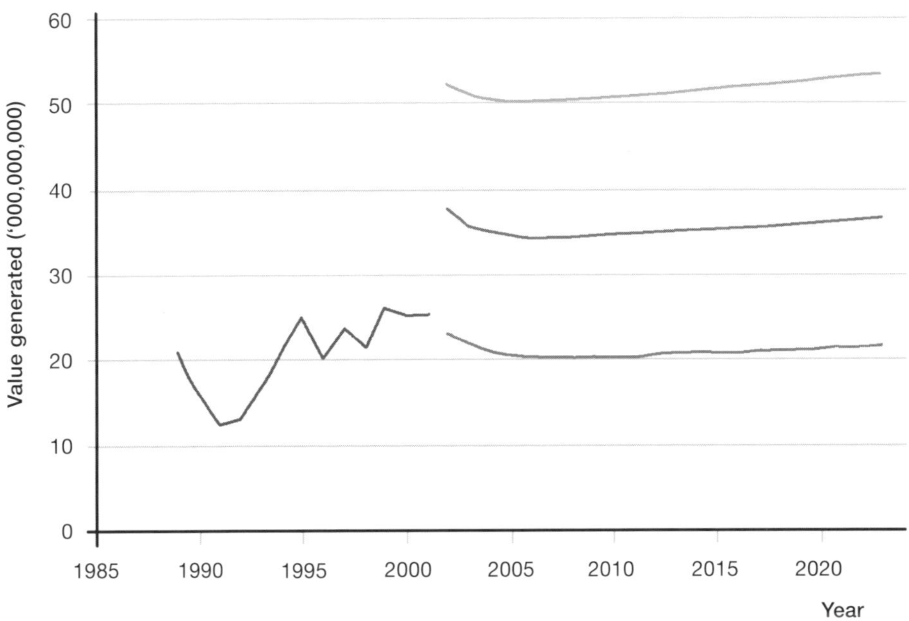

The stated strategies in the context of General Motors historical performance forecast a future of slightly declining volumes and value generation. The value of the 'environment' variable θ = -1.56 billion suggests that GM's competitive environment is robbing it of value. The net present value of forecasts is \$36 billion. With 560 million shares outstanding, this implies a share price of \$65 (with 90% confidence limits of \$37 and \$94 per share) close to the \$30 to \$70 range in which GM traded in 2002. Note that this difference could reflect overly optimistic assumptions or the failure of the Financial Dynamics forecasts to incorporate all of the existing competitive pressures; conversely, the difference could reflect the failure of audited financials to fully value GM's strategy and operations in the stock market. Financial Dynamics gives the analyst a vehicle with which to identify additional data collection and analyses which can seek to explain current market valuations.

Consider how the Value Maps might be expanded in pursuit of a model with greater resolution. Following the lead of GM's own financial statements in the past few years, we might break out the finance operations from the automotive products.

To refine the value map, we can split GM into its auto production and finance arms as shown in figure 10.8.

Figure 10.8 GM's Value Map with Financing

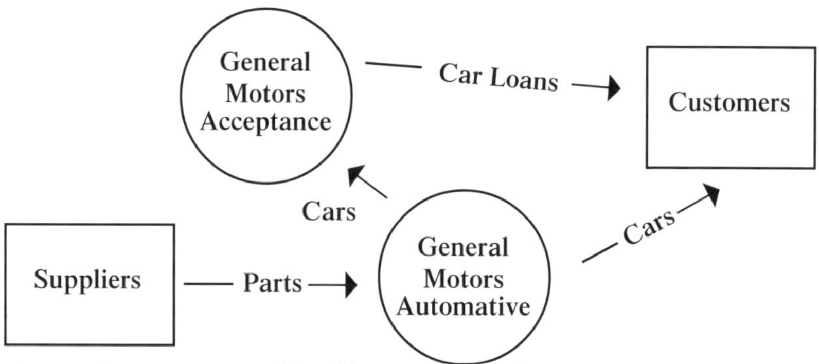

Further refinements would add suppliers who are mainly funneled through Covisint, the auto industry's captive e-commerce market; as shown in figure 10.9.

Figure 10.9 GM's Value Map with Financing and Sourcing

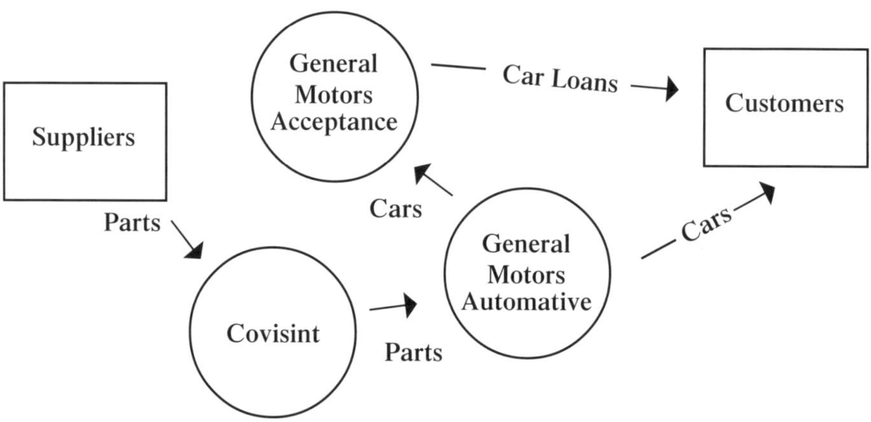

For auto industry analysts, this is probably nothing new; certainly they are evaluating the industry at this level of detail. What Financial Dynamics does is to formalize the process by which values are obtained so that the same discipline can be applied to less well tracked firms, or even to projects and private ventures which are not tracked in the stock market.

Why is it important to adopt a more complex Financial Dynamics approach to valuation?

Accounting was formalized over six centuries ago, long before computers (or for that matter even widespread use of the abacus). Discounted cash flow (DCF) was formalized in the 1930s; the capital asset pricing model (CAPM) (useful for determining equity capital costs) in the 1950s. They all invoke simplifications or antiquated models that incorporate calculation shortcuts, necessary for application before computers were widely available, and user friendly.

Today, we keep more power on our desktops than the whole company had two decades ago; and it is linked into a global database, with real-time search tools that were unimaginable only a decade ago. We can do casual analyses using modern tools from econometrics that inject consistency and explicit definition into the valuation process.

[1] Quoted in "A Survey of International Finance," *The Economist*, May 18, 2002, p. 18.

[2] Introduced May 14, 2002

[3] The caret ^ over a variable indicates that this is an estimator – a value estimated by regressing the model against historical data.

[4] W. Baumol, *The Free-Market Innovation Machine* (Princeton, N.J.: Princeton University Press, 2002).

[5] Source: Economy.com and Business Week, 13 Jan 2003 p. 74-5

[6] Ibid.

[7] "Auto Stocks Drop on Analyst Report," *AP Online*, June 20, 2002 reports that Morgan Stanley analyst Stephen Girsky cited the likelihood that sales would level off in a tight pricing environment and doubts about the auto-makers' efforts to reduce manufacturing capacity. Lower sales in 2003 suggest that the probability of significant earnings growth in 2003 is low.

[8] The company took the name General Motors, because its first choice, United Motors, had already been taken by a firm domiciled in Delaware, where it incorporated.

[9] Homologation is the part of R&D that ensures the production vehicles meet regulatory standards, and contributes around 5–7% of new car cost.

[10] Materials cost in a typical car is around half that required in the 1970s. Indeed, the cost of raw materials is the main reason that older auto designs such as the original Volkswagen Beetle and the Austin Mini became unprofitable, even though a robust market existed for the products. Their modern reincarnations are lighter and much less expensive to make than the originals, making them once again profitable.

[11] Boston Consulting Group.

[12] Center for Automotive Research at the University of Michigan, U.S. Department of Commerce, and Accenture.

[13] Sources: Company reports; *Automotive News*

[14] Andy Kjellgren, "Segway Hits Bump in its Progress," *The Atlanta Journal-Constitution*, March 5, 2002.

[15] Stanley Hart and Alvin Spivak, "The Elephant in the Bedroom: Automobile Dependence and Denial." New York: Hope Publishing House, 1993; The General Accounting Office cites estimated U.S. national productivity losses from congestion of US$100 billion annually, and truck-delay costs from congestion of US$24 to US$40 billion per year

[16] James MacKenzie, Roger Dower & Donald Chen; The Going Rate: What It Really Costs To Drive; Washington DC: World Resources Institute, 20006 ,p. 18–24.

[17] "Global Pile-up," *The Economist*, May 8, 1997.

[18] At that time in the early 17th century, only two engines existed – the water-wheel and the windmill. Both were huge, and yielded a maximum of 10 horsepower.

[19] Despite his popularity as a "business guru," Taylor continued to provide such services at his customary fee of US$35 per day.

[20] "The Art of Overtaking," *The Economist*, September 6, 2001.

[21] A precise calculation is more complex, but at the 95% confidence limit on either the P or the Q parameter, the approximation I have applied here is accurate enough given the data. If the reader needs to pursue greater accuracy, then a convolution formula must be used to multiply the $\tilde{P}$ and $\tilde{Q}$ random variables. Given the small number of datapoints, and the number of assumptions in forecasting, the accuracy is satisfactory.

CHAPTER 11

The Dark Side of Accounting

I am deeply concerned about some of the accounting practices that take place in America.

– George W. Bush, U.S. President[1]

I've been tracking this matter for a lifetime, and my greatest surprise was the sheer scale of the inadequacy of the accounting profession and some of its most prominent members.

– John Kenneth Galbraith[1]

The profession of accounting is, in fact, in crisis.

– Paul A. Volcker[2]

Why is there auditing? Oddly enough, the AICPA, FASB and SEC are willing to go to great length (1) to discuss what activities constitute an audit, and (2) to describe what an audit *is not. But,* they fail to describe what an audit *is*. Accountants turn coy when confronted with impolite questions such as '*what purpose does an audit serve?*' Expect the inquirer of such an embarrassing question to be given a condescending nod usually reserved for those who just don't get it.

Chapter 2 presented a short history of accounting where I noted that modern auditing was initiated in 1934 in the aftermath of the 1929 stockmarket crash. There was little precedent for audits at that time; legislation and methodology was implemented under intense political scrutiny, and was rushed into service with little consideration for the eventual form or purpose of the audit. It is interesting to note parallels with the politics and urgency following passage of the Sarbanes-Oxley Act of 2002, which at the time of this writing is still being interpreted.

In a world where time is money, analysts and investors suffer unconscionable delays waiting for an audit opinion. Transaction summaries may sit around unaudited for up to a year and a half before becoming 'audited account values' – a lifetime in 21st century businesses. The question 'Why is there auditing?' is anything but rhetorical. Answering it should be relatively easy. Right? The audit is a process or set of activities that are applied to a given subset of any publicly traded firm's annual financial statements. The audit process (by definition) converts them from 'unaudited' to 'audited' financial statements. We only need to ask 'What is the difference between the two, and is it worth the wait?' Answering this requires that we take a deeper look into the legal and operational specifics of auditing.

Professor Donald Langevoort of Georgetown Law School, an expert on securities audit regulation, identifies two perspectives on the purpose of auditing.[2] The narrow perspective is that the audit is no more than a contractual agreement between a company and its auditor, and that audit standards are subject to negotiation. This may indeed be the perspective that lawyers assume when defending auditors in court. But a more mainstream perspective, observes Langevoort, is that accounting standards should be uniform across all companies, and the audit is the mechanism for assuring this uniformity, assuring comparability of financial statements. The audit opinion itself promises only three things: (1) the accounts follow GAAP, thus presumably are consistent between companies; (2) the accounts are 'fairly' stated, whatever that means, and (3) GAAP is applied consistently over time. Much of the current debate over accounting stems from a shared belief that the body of rules that comprise GAAP are irrelevant to contemporary business, too loosely defined and sometimes misleading (as in the case, for example, of accounting for stock options).

The deeper, and typically unasked question is whether 'consistency' is the objective investors crave. Or would they rather have *accuracy*, *completeness,* and *usefulness in decisions.* Emerson famously observed that "A foolish consistency is the hobgoblin of little minds, adored by little statesmen and philosophers and divines." Consistency is safe but not very useful. And it gets us into trouble in ways that aren't initially obvious.

This is the 'dark side of accounting.' The phrase is a double entendre: on the one hand, it depicts something dark, malicious, and sinister; on the other, it signifies the unilluminated or unrevealed part of something. Before there was gaslight, or other wide-area nocturnal illumination, the dark was itself a source of insecurity. Our vision is at its weakest in the dark, and so important is it to us that we can't be blamed for jumping at every sound when it is impaired. As such, it has become a metaphor for that which is not understood, not trusted, and what is most definitely feared.

The "dark side" is an enduring theme in modern literature. My cursory search of bookseller Amazon yielded 217 titles that contain the words "dark side," including *The Dark Side of the All-American Meal*, *The Dark Side of Christian History*, *The Dark Side Sourcebook*, *The Dark Side of Leadership*, and, of course, *The Dark Side of Valuation*. No field of human endeavor is impervious to the dark side. Recent accounting scandals have made us painfully aware of the dark side of accounting, even though we may have trouble putting our finger on exactly what is wrong with the profession. This book, and my previous book *Valuing Technology*, have argued that the dark side arises not from what accounting does, but rather from what it fails to do. And its failures have grown significantly more damaging by the 21st century. Economist John Maynard Keynes couched it eloquently in saying that "the difficulty lies, not in the new ideas, but in escaping from the old ones, which ramify, for those brought up as most of us have been, into every corner of our minds."[3] George Lucas,[4] who popularized the term through his *Star Wars* epics, has suggested a common source for the dark side: "the truths we cling to depend greatly on our own point of view."

Accounting's dark side surely has *its* source in the "truths" that have held the profession captive for six centuries. Accounting statements and accounting practice have diverged so significantly from the reality of business that they have led managers and investors into destructive (and often self-destructive) behavior. The first reference I recall to "the dark side of accounting" was in a speech to the American Institute of Certified Public Accountants in 1998 by Walter P. Schuetze, then chief accountant of the Securities and Exchange Commission. He was referring to shortcomings in financial accounting that hampered the SEC's enforcement of securities regulations. Bias, lack of useful information, excesses of purposeless information, incompleteness, and other shortcomings of generally accepted accounting principles (GAAP) were wreaking havoc with the SEC's enforcement efforts during the peak of the dot-com bubble.

Certainly, there has been sufficient debate in the popular press on audit quality since the Enron and WorldCom debacles. It is unlikely that the audit profession will survive the next decade without major reform. But reform is unlikely in the light of the intense conservatism of the four largest audit firms – especially in the face of increasing liability and the risk of lawsuits. Any new language will most likely be provided from an outside source. Credit rating agencies such as Moody's and Standard & Poor's are positioning themselves to provide this information. But it is more likely that a complete break may be needed for a new language appropriate to the knowledge economy to evolve.

The field of accounting aided and abetted some spectacular corporate failures and investor scams in the 1990s. No doubt the most egregious was the fall of the once-revered Andersen, where failure was closely identified with the abuses at energy giant Enron. What is easy to forget is that the Enron failure was preceded by a string of other failures at Andersen, all victims of the inability of their accounting and auditing practice to come to grips with the post-industrial economy. No less an authority than Joseph Berardino, head of Andersen during the Enron debacle, complained, "Our financial reporting model is broken."[5] Even more damning is the reaction of the investment community, reflected in the comment of Robert Willens, Lehman Brothers Inc. accounting expert: "I don't know anyone who uses GAAP net income anymore for anything."[6]

ACCOUNTING, THEN AND NOW[7]

On May 14, 2002, in a bid to come up with rules to unify earnings and find a better measure for operating performance, Standard & Poor's introduced the concept of "core earnings," a measure aimed at providing a standard, comparable number that most closely reflects the ongoing earnings that a company produces from its operations. It contrasts with both pro forma figures and the earnings number calculated according to GAAP, because S&P includes many costs that companies and Wall Street analysts would rather exclude, including expenses from stock option grants and the restructuring of ongoing businesses, as well as the writing down of the value of certain assets, pension costs, and research and development purchased from another party. At the same time, core earnings exclude income that it doesn't consider as coming from a company's operating business, such as gains on its pension plan, asset sales, or hedging operations, or such things as income from litigation or insurance.

The accounting profession has been aware of these shortcomings for at least two decades, but has opposed change in order to hold on to old ways. Several alternative organizations are waiting in the wings to take on the audit and standards setting from accounting. Since the 1930s, the accounting profession has focused on the needs of shareholders. Banks, on the other hand, have looked to *credit rating* agencies, such as John Moody's *Manual of Railroad Securities*, first published in 1909, or the Standard Co. bond grading in 1916, or Poor's and Fitch reports on corporate debt in the 1920s.

Inside the growing industrial firms – automobiles, ships, petroleum – at the turn of the century were the *cost accountants*, whose clientele were corporate managers. Focus on discretionary budgets in the 1950s brought forth the discipline called *managerial accounting*.[8] Out of managerial accounting at giants such as Procter & Gamble and General Electric came techniques that later found their way into various proprietary methods. The most successful of these proprietary methods has been Stern Stewart's economic value added (EVA) using a variation on the *residual income method*. Other consulting firms are promoting their own proprietary valuation methods – HOLT Value Associates has cash-flow return on investment; Boston Consulting Group has total business return (TBR) and cash flow return on investment (CFROI); LEK/Alcar has shareholder value added (SVA); Marakon Associates has economic profits (EP); and KPMG has economic value management (EVM). Each group imposed its own interests on "value" – credit raters on ability to repay debt; cost accountants on the dynamics of the production process.

In another area, forensic accounting is a growing business. Perhaps the best-known firm selling research based on forensic audits is Dr. Howard Schilit's consultancy, the Rockville, Maryland-based Center for Financial Research and Analysis (CFRA) founded in 1994. Schilit correctly alerted subscribers to problems at Enron, Vivendi, Xerox, and MicroStrategy well in advance of actual fraud proceedings. Even if uniformly accurate valuations and accounting are impossible, there will always be a market to alert investors about problems early.

These extensions and revisions to accounting attempt to adjust to the dictates of a changed world. Classical accounting systems work specifies a limited set of transaction processing cycles, which define the business tasks of a firm and are responsible for the generation of account balances. In the conduct of research and the teaching of accounting, the concept of transaction processing cycles is generally covered under the rubric of accounting information systems (AIS). AIS is generally concerned with building computer systems. Accounting systems represented the first commercially important applications of computers, and developed rapidly after IBM's introduction of its System 360 in April 1964. The study of accounting information systems only superficially touches on technology, and devotes most of its focus to describing and flowcharting the processes by which account balances are generated. The account balances are the "stocks" of wealth that a firm acquires and spends; while the transaction processing cycles are the "flows" of wealth that add to or deplete these "stocks."

Accounting firms, in their annual audit of a firm, must ensure that account balances are "fairly stated in accordance with GAAP" and must also include a separate letter in the audit papers that attests to the proper functioning of the internal control and transaction processing systems. Problems in either can result in qualification of the financial statements. Most firms audit the account balances by a process of taking inventories of originating transaction documents that support the account balance. At times, certain firms – for example, Arthur Andersen – have audited the transaction flows (a much more expensive and labor-intensive task).

Transaction processing has in the past generally been classified into five cycles, which tend to be most applicable to agricultural or manufacturing operations. These classical transaction processing cycles are:

- *The revenue cycle:* This includes sales, collections on accounts receivable, and other events related to the distribution of goods and services.
- *The expenditure cycle:* This includes disbursements for inventory purchase and accounts payable retirement, cost of goods sold, and other events related to the acquisition of goods and services.
- *The inventory production cycle:* Events related to the transformation of resources into goods and services.

- *The finance cycle:* Events related to the acquisition and management of capital funds, including cash.
- *The payroll cycle:* Events related to the deployment of personnel (in practice, these systems combine quantitative accounting with qualitative personnel assessments and histories).

These five cycles are most applicable to a manufacturing company that buys raw materials and converts them to salable inventory for profit. They tend not to be applicable to knowledge firms such as Dreamworks, Microsoft, and Bloomberg, which generate value but not inventory in the traditional sense. Thus, a more general definition of transaction cycles is needed for valuing knowledge economy firms.

These five cycles did not spring fully fledged, but evolved through a long and ancient sequence of innovations. Often the reasons for the innovations have been lost, but the innovations themselves have carried on as "memes" – the ideas and other elements of culture that tend to make copies of themselves and are self-replicating like genes; that are stored in human brains and passed on by imitation. Accounting systems are, and in the past have been, designed mainly to provide information that managers and third parties can use in decision-making. They also serve other purposes: to produce operating documents, to protect the company's assets, to provide data for company tax returns, and, in some cases, to provide the basis for reimbursement of costs by clients or customers. The nature and variety of objectives has changed significantly over history.

I argued in chapter 2 that accounting has been in continual transition since the time of King Croessus of Lydia. The days are numbered for that particular brand of financial accounting given legal credence by the Securities Exchange Act of 1934. It will either evolve or be replaced by tools more relevant to the knowledge economy. But what about auditing? Whenever I suggest that auditing's days may be number, I am greeted with incredulous stares – 'what could take its place?' This is certainly a question worth pursuing.

One school suggests that auditing is less important for what it does, than for the discipline it imposes on the firm and financial markets. The threat of a qualified opinion – goes this school – is enough to motivate firms to clean up their accounting systems, and report reasonably reliable numbers. The flaw, though, is that since most of what the market values seems not to be covered in the audit (because of the narrow scope of GAAP, which ignores most intangible assets), and firms know this, then they actually are motivated to shift their game playing into whatever arena is not covered by GAAP. Whenever GAAP threatens to invade their gaming arena, lobbyists put pressure to maintain the status quo.

Like accounting, auditing must evolve. The volume of recorded transactions is today several million times that of the 1930s, and cannot be effectively audited without some degree of robotics and automation. Nor can auditors afford to ignore advances in statistics which can guide auditors in cost effective sampling and decision making. Both are used inside firms to provide management with the information they need. The evolution of auditing is more a matter of will than of technology; the technology is available once the political, legal and educational environments favor advancement of the practice of auditing.

THE ACCOUNTING PROFESSION'S RAISON D'ÊTRE

Accounting has historically been designed to track value, at the request of owners and others who have legal rights to property. Today, though, the main justification for the existence of the four major accounting firms is auditing – that is, the verification that accounting for publicly traded companies is fair, consistent, and complies with GAAP. During the 1980s, the AICPA's repeal of the *Code of Professional Conduct*'s sections 502-1, 502-3, and 502-4 allowed auditors to advertise and solicit business, exposing them to the full blast of market forces. Auditing, like any other business, must control costs to ensure they do not exceed revenues. This became increasingly difficult in the 1980s' environment of price competition for audit business at the same time that audits grew substantially more complex due to rapid moves by their clients into knowledge-intensive products and operations. Consulting services were aggressively expanded to cover the shortfall in revenues. Audits were priced as "loss-leaders" in order for the firm to acquire the business associated with the design and implementation of computer-based accounting systems which might be recommended in the course of the audit. This not only undermined the independence of the audit, but essentially transferred economic power into the hands of systems consultants who suffered none of the ethical constraints of auditors. The resulting tensions over revenue sharing erupted first in the acrimonious split between Andersen and their consulting arm, and later in clashes at the other (then) Big 8 firms.

The audit function is very much a 20th-century development in the history of accounting. At the beginning of the 20th century, the rapid increase in the size of American companies created a need for audits based on selected tests of items constituting account balances or classes of transactions – what we have now come to know as "sampling."

Sampling allows an audit to be performed cost-effectively. The largest expenditure of audit effort is typically directed towards verifying inventory and accounts receivable. Audit samples choose a subset of transactions. Each transaction in this subset is, in theory, closely inspected to ensure that it indeed has been properly recorded and classified by the firm's accounting systems (using what auditors call "internal control systems tests") and that it is properly incorporated into the published financial reports (using what auditors call "substantive tests"). There is nothing magical about the audit process – the same rules that apply in statistical decision analysis apply to audits. Failure to randomly select samples for testing (which ensures that every transaction is potentially represented in the audit) and to properly draw inferences from these tests about the entire set of financial statements undermines the validity of the audit tests. Statistical approaches to sampling and inference have been constructed precisely to ensure that human bias, systematic omissions, and other foibles do not influence the conduct of the tests. Unfortunately, this fact has been met with controversy in the profession, dating back to the beginning of the 20th century.

The earliest references to audit sampling appear in a program of audit procedures printed in 1917 by the U.S. Federal Reserve, and contain vague recommendations such as selecting "a few book items" of inventory. Nothing much more was said to refine the practice of auditing until 1955, when the American Institute of Accountants (later to become the AICPA) published *A Case Study of the Extent of Audit Samples*, which summarized audit programs prepared by several CPAs to indicate the extent of audit sampling each considered necessary.

The first pronouncement on the subject of statistical sampling in auditing was a special report, *Statistical Sampling and the Independent Auditor*, issued by the AICPA in 1962. The report concluded that statistical sampling was permitted under generally accepted auditing standards (GAAS). A second report, *Relationship of Statistical Sampling to Generally Accepted Auditing Standards*, issued in 1964, was later included as appendix A of *Statement on Auditing Procedure (SAP) 54: The Auditor's Study and Evaluation of Internal Control* which was, again, reused later as AU section 320 of *Statement on Auditing Standards (SAS) 1: Codification of Auditing Standards and Procedures.* Some of that material was extended in 1972 as Appendix B of SAP 54. There were, in addition, Statements on Auditing Procedure (SAP) 33 and 36,which gave tepid support to the use of statistics, but in general encouraged the continuation of what the practice called judgmental sampling – an *ad hoc* selection of samples that suffered from the biases and foibles that statistics were designed to control. In 1981, the AICPA rolled all of this into a single standard SAS 39, titled *Audit Sampling.* Apparently, the AICPA felt that there was very little in the way of new audit developments over the prior three decades, as SAS 39 more or less repeats their position in the mid-1960s. Several pronouncements have been released since that deal with the application of statistical sampling, but not the performance of tests, sampling, or inferences based on these. For example, SAS 47 discusses auditor risk and the murky concept of materiality (the auditor's dispersion or error statistic) in planning an audit. SAS 55 and SAS 78 address internal control and control risk, but fail to provide objective procedures or formulas for assessing either. The lack of formal, objective procedures to quantify concepts such as materiality, audit risk, and internal control risk invites inconsistent auditing, and major errors such as the failure to identify bogus revenue at WorldCom. This also should strike you as a bit ironic in a profession that has defended outdated reporting concepts based on their principle of "objectivity."

THE TROUBLE WITH AUDIT SAMPLING

So, what do the auditors actually do during an audit? We have an indication from the AICPA in its official publications; specifically, *SAS 39: Audit Sampling*, and the book *Audit Sampling*, which provide tables of suggested sample sizes dependent on various parameters. Along with this, the AICPA has concocted its own lexicon of jargon, often with vague definitions that may give latitude for legal maneuvering. In fact, they recommend "auditors may use whatever terms they prefer as long as they understand the relationship of those terms to the concepts in SAS 39."[11] This certainly seems an invitation to confusion, as it divorces audit sampling from the large, learned, and scientific body of statistical literature. It seeds misconceptions among auditors who already have difficulty with the concepts. This professional proliferation of neologisms does nothing to control actual misstatement of accounts, and has contributed greatly to the audit profession's troubles. For example, Chapter 1 of *Audit Sampling* recommends the definitions set out in Table 11.1 in place of standard statistical usage:[12]

Table 11.1 Confusing Terminology

Standard Term	SAS 39 Term
α or Neyman-Pearson type I risk	• risk of assessing control risk too high • risk of incorrect rejection
β or Neyman-Pearson type II risk	• risk of assessing control risk too low • risk of incorrect acceptance
Precision (the inverse of variance)	• allowance for sampling risk
Risk	• complement of reliability • complement of confidence level
Maximum allowable error in an account	• tolerable misstatement • materiality

But the problems extend beyond just terminology. The AICPA gets the numbers wrong, at least in the selection of sample sizes and inference of risk and materiality.

The default mode for statistical sampling goes by a quaint name, "discovery sampling," which means doing just enough work to ensure that the audit might probably discover an error transaction if it existed. Unfortunately, discovery sampling in the AICPA's Statement of Auditing Standards means that you *assume to begin with that there are no errors at all in the client's transactions*. The AICPA provides no explanation as to why one would ever make such an implausible assumption. One might argue that placing naïve and credulous faith in the client's accounting systems, without first having performed an audit, is in fact in contradiction with the implicit function of the auditor – that is, to be skeptical. The advantage of such an assumption is, of course, that the resulting sample sizes are commensurately small, and thus the audit testing will be cheap. But the probability of finding errors will also be commensurately small unless there is nearly a complete breakdown in the client's accounting systems. In truth, if errors were found in the prior year's audit, it would be very difficult to justify the assumptions underlying discovery sampling.

Let's look more closely at the sampling problem confronting auditors. The profession's articulation of the audit sampling problem seems straightforward enough: tests must assure that *material errors* have not occurred in the audited account balances. This may be restated mathematically:

$$P\,(|\hat{\theta} - \theta\,| > M) < \alpha$$

where:

P (·) is the probability of the event in the brackets;

$\hat{\theta}$ and θ are the estimate and actual value of the account, the estimate obviously being the product of the auditor's audit;

M is the materiality level (that is, the maximum error allowed in the account balance for which the auditor would still say that the account is "*fairly*" stated); and

α is the risk that the account is materially in error (also called Type I (Neyman-Pearson) risk).

Now consider the implications of this formulation for a normal distribution. The auditor will assume that $\hat{\theta}$ is an unbiased, normally distributed estimator of θ; let it have population mean μ and variance σ^2. Then the two-sided test criterion can be restated:

Choose sample size n, large enough to make $P\,(|\hat{\theta} - \theta\,| > z_{\alpha/2}\sigma) = \alpha$.

For a population of N, the criterion becomes $z_{\alpha/2}\sqrt{\frac{\sigma^2}{n}}\left(\frac{N-n}{N}\right) = M$ or $n = \left(\frac{M^2}{\sigma^2(z_{\alpha/2})^2} + \frac{1}{N}\right)^{-1} \rightarrow \frac{\sigma^2\,(z_{\alpha/2})^2}{M^2}$ as the population size grows large (where $z_{\alpha/2}$ is the confidence bound of a $N(0,1)$ distribution). Now, instead of some predetermined value of M, audit "fairness" is in practice more likely to dictate a "relative" materiality level r (that is, a proportion) such that $r\theta = M$. In this case, $\frac{\sigma^2(z_{\alpha/2})^2}{M^2} = \left(\frac{\sigma z_{\alpha/2}}{r\theta}\right)^2 = \left(\frac{\gamma z_{\alpha/2}}{r}\right)^2$ where $\gamma = \frac{\sigma}{\theta}$ is the coefficient of variation. I say this because where auditors' materiality choices have been revealed in court proceedings, they have usually been stated as percentages – that is, in relative terms.

Let's look at two situations that the auditor might encounter, and consider how discovery sampling might lead the auditor astray. First assume that a prior audit sample of 10 transactions found one that was counted twice in the final accounts (something like the result of a "dark swap" at WorldCom or Enron, where a sales transaction of bandwidth might have been counted several times). Assume that this was the only error. Then, in the current audit's sample selection, the presumption of "no errors in the population of client's transactions" is not justified. In fact, the best estimate you have is that the quantity of transactions is 110% of its true value – that is, in estimating value P and volume Q, you might assume no error in P, but you should assume an overstatement of 10% in Q and thus that the true value of the account θ is 90% of the stated value. Now, this problem is likely to require a much larger sample size because $|\hat{\theta} - \theta|$ will grow large as the value moves toward the center of the density function. It is more probable that there is a material error, and harder to assert that $P(|\hat{\theta} - \theta| > M)$ is less than α.

The second situation involves the assumption of multiple business processes, which is normal practice in large firms. Suppose that your automotive firm makes vehicles for two markets: the first is a passenger car that sells for US$10,000; the second is a military vehicle that sells for US$100,000. The Financial Dynamics approach would separate these two business processes in the value map, and deal with the estimation of each separately. Accounting firms audit the account, which in this case would be sales, thus combining the sales from both processes. Table 11.2 sets out some examples of transaction statistics for a given year.

Table 11.2 An Example of Mixed Sales Distributions

	Value/Transaction		No. of Transactions/Year	
	Mean	Std Deviation	Mean	Std Deviation
Passenger car	10	10	1000	1000
Military vehicle	100	100	10	10

Sampling without segregation around the two business processes will yield a distribution that appears in Figure 11.1.

Figure 11.1 Assumed distribution of audit sample if auditors do not segregate business processes

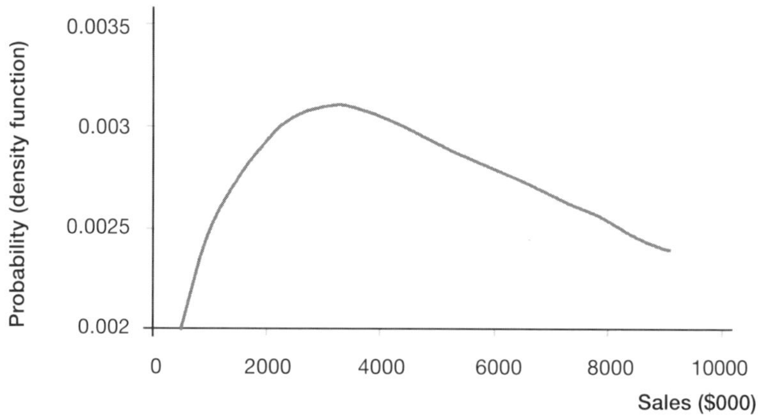

This distribution has its mode at US$3.4 million, its mean at US$11 million, its standard deviation of US$11 million, and is right skewed.[13] Assume (consistent with AICPA's default audit assumption) that both the value and number of transactions are normally distributed with the parameters shown in Table 11.2.

Figure 11.2 True distribution of audit sample (two processes)

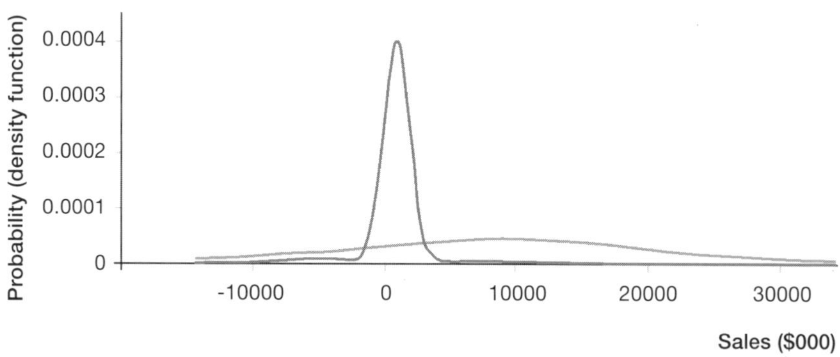

Now assume that materiality is set to a relative error of $r = 10\%$. The $n = \frac{\gamma^2 (z_{\alpha/2})^2}{r^2}$ where $\gamma = \frac{\sigma}{\theta}$ is the coefficient of variation. In Figure 11.1, with $\alpha = 10\%$ the sample size that should be taken according to the AICPA's formula is $n = \frac{(^{11}/11)^2 \text{ x } 1.65}{.1^2} = 165$. In contrast, for Figure 11.2, two samples would be taken, where the materiality is the same for both – US\$1.1 million. $n = \frac{(^{10}/10)^2 \text{ x } 1.65}{.11^2} \cong 136$ and $n = \frac{(^{1}/1)^2 \text{ x } 1.65}{1.1^2} \cong 1$ for a total sample size of 137. Segregating the two processes and auditing each separately is obviously the more efficient of the two approaches.

It should be noted that the AICPA does recommend stratification of samples, but the stratification they recommend is on transaction size in the account, not on the underlying business process that generated it. This is because the focus is placed on auditing the account balances in the financial statements; not on the processes that generate them. By the time information is aggregated in the accounts, their process history has been lost to the accounting system.

The damage done to the audit profession by these inefficiencies is enormous. The profession must endure charges of negligence and corruption when, in fact, the guidance they receive from their professional bodies is flawed. Investors are highly sensitive to a flawed audit function. An auditor's report is like Consumers' Report for products sold in the stock market. It wouldn't be surprising to find consumers express anger if Consumers' Reports lied about the reliability of cars, because the auto manufacturers were paying them a retainer. Unfortunately, this is the situation that auditors now find themselves in, especially since they are making most of their revenue from consulting, rather than auditing.

ACCOUNTING EXTENSION FOR THE KNOWLEDGE ECONOMY

Can Financial Dynamics be applied to solve some of accounting's most pressing current problems? I believe it can, and in doing so, will follow in the tradition set by prior extensions to financial accounting – debt valuation, cost accounting, managerial accounting, and various "boutique" metrics – of first finding use as a specialized valuation method, but over time will be incorporated into accounting systems, and ultimately the practice of financial accounting.

Fiscal archeology with Financial Dynamics offers another alternative as a stopgap approach until fundamental new forms of reporting are available in the marketplace for financial information. After all, until 1933, accounting firms sold their audit services to firms in the open market, as did Moody's and S&P.

Because Financial Dynamics bases value flows on appropriate monetary or non-monetary transaction "quanta" and strategic "value drivers," it offers an alternative to the aggregation of firm activities in monetary units for expressing the results of firm operations. As we mentioned earlier, the problems with monetary units occur in four areas:

1. *Inflation/deflation*: They are unstable over time.
2. *Erratic exchange rates*: They are unstable across global operations.
3. *Market failures*: They fail to capture value where markets are either thinly traded or nonexistent
4. *Rapid depreciation*: Historical cost fails to properly assess value where there is rapid depreciation of goods. This is especially a problem where technology is accelerating rapidly.

At its root, accounting is and always has been the process of recording the value of transactions and summarizing them for the benefit of some group of users. Differences arise in the way that the real world is simplified in building the business model. Traditional accounting has adopted a model suitable for agriculture and commodities trading, but incomplete in tracking transactions in the industrial and knowledge economies. Only economic events that involve "arms-length transactions" – that is, those with parties external to the firm – are included in the accounts. As management grew more complex, cost and managerial accounting were adopted by industrial firms for decisions involving specific transactions inside the firm (production inventory, and discretionary budget transactions, respectively). Meanwhile, stakeholders other than the company's owners were addressed by bond-rating firms such as Moody's and S&P. Each advance came about in response to new complexities in business, and new groups of stakeholders.

The knowledge economy knocks business up to a new level of complexity. It has unearthed new stakeholders, often who only gain recourse to company management through litigation, regulation, and politics. Two episodes in recent financial history have brought to light new stakeholders. After the "junk bond"-fueled crash of 1987, depositors in failed Savings & Loan institutions found that they needed more accountability than originally assumed. After the dot-com-fueled crash of 2000, employees and middle-class investors similarly demanded tighter accountability.

Each new constituency requires a rethinking of the way that we track, record, and report on corporate transactions. Accounting will still be focused on wealth and value; but this is likely to be interpreted differently by different stakeholders. An economist would define income as the change in the company's wealth or welfare during a period of time, from all sources other than the injection or withdrawal of investment funds. Accountants generally have in the past rejected this approach, because such a measure would rely too much on estimates of what will happen in the future, estimates that would not be readily susceptible to independent verification. Instead, accountants have adopted a *transactions approach* to income measurement. They recognize as income only those increases in wealth that can be substantiated from data pertaining to actual transactions that have taken place with persons outside the company. Yet, this approach to accounting has been criticized on the grounds that it confuses monetary measures with the underlying realities when the prices of many goods and services have been changing rapidly, or when value is interpreted differently by different stakeholders.

In *Valuing Technology*, I suggested a set of 10 principles that should be added to accounting's existing principles in order to update accounting to effectively deal with the knowledge economy. At this point, it may be worthwhile to see if the methods of Financial Dynamics might suitably be applied to extend accounting's process of recording the value of transactions and summarizing them for the benefit of some group of users.

Financial Dynamics' suitability will be determined by two aspects of its application – first to transaction recording, and second to reporting to stakeholders.

TRANSACTION RECORDING

In traditional accounting systems, transactions are first recorded to a journal. (The accounting terminology for a transaction is a "journal entry.") This journal may apply to single accounts, such as a Sales Journal, or may record multiple accounts, as does the General Journal. Summaries may be continuously reported, as they are in perpetual inventory systems, or in "flash" sales. Periodically, these totals are "audited" – the firm's term for reviewing and finalizing the journal entries prior to sending them on for summarization on the trial balance, and reporting on the published financial statements.

In Financial Dynamics, the journals record the transactions and use them to forecast a stream of future value flows in a dynamic context. Following the mathematical appendix to Chapter 6, the estimation model for either volume or transaction value is (where either is represented by Y in the formulas presented in Chapter 6 Appendix B)

$$Y_t = \sum f_i(X_{i,t}) + \theta + u_t(\rho)$$

where the Xs are *strategy drivers* – that is, the factors that are actively managed in the firm's strategy; θ represents other contemporaneous components of value not influenced by managerial strategy; ρ represents strategy "lag" – that is, the autocorrelation of current value generation with the prior period value creation, where the error terms u_t follow a first-order autoregressive model $u_t = \rho u_{t-1} + \varepsilon_t$ where autocorrelation $|\rho| < 1$ and the ε_t are independent with zero means and independent of the $X_{i,t}$. The *strategy drivers* are proactively set by management, and the estimator $\hat{\theta}$ that solves $\min_\theta \sum_{t=1}^{T}\left(Y_t - \sum_i f_i(X_{i,T+1}) - \theta\right)^2$ along with residual autocorrelation ρ_T captures the uncertainty in the forecast.

Once these are complete, the prediction of Y_{T+1} is given by

$$\hat{Y}_{T+1} = \hat{\rho}_T\left(Y_t - \sum_i f_i(X_{i,T}) - \hat{\theta}_T\right) + \sum_i f_i(X_{i,T+1}) - \hat{\theta}_T$$

More generally, for any forecasting horizon h, we can show by extending this exercise that

$$E_{\theta,\rho}(Y_{T+h} \mid Y_1,\ldots,Y_T) = \sum_i f_i(X_{i,T+h}) + \theta + \rho^h\left(Y_T - \sum_i f_i(X_{i,T}) + \theta\right)$$

Thus, a prediction of Y_{T+h} depends on the one-period strategy lag $\hat{\rho}_T$, and the strategy driver values at $T + h$ as follows (formulated with all of the transactions up to time T).

$$\hat{Y}_{T+h} = (\hat{\rho}_T)^h \left(Y_T - \sum_i f_i(\mathrm{X}_{i,T}) - \hat{\theta}_T\right) + \sum_i f_i(\mathrm{X}_{i,T+1}) + \hat{\theta}_T$$

Now assume that a new transaction arrives. The value P and volume Q (generically represented here as Y) can be updated as:

$$_{T+1}\hat{Y}_{T+h} = (\hat{\rho}_T)^{h-1} \left(Y_{T+1} - \sum_i f_i(\mathrm{X}_{i,T+1}) - \hat{\theta}_T\right) + \sum_i f_i(\mathrm{X}_{i,T+h}) + \hat{\theta}_T.$$

Note that this *does not* require a recalculation of $\hat{\theta}_T$ or $\hat{\rho}_T$, nor new assessments of $\sum_i f_i(\mathrm{X}_{i,T+1})$; thus, it has all the computational simplicity required of a journal entry. It can be thought of as the Financial Dynamics counterpart to a perpetual inventory. Perpetual inventory, though, needs to be periodically audited with a "periodic count" of inventory to tie out the account balance to the real world. The Financial Dynamics counterpart to this "audit" is an updating every $m < h$ periods of $\hat{\theta}_T$ and $\hat{\rho}_T$ to $\hat{\theta}_{T+m}$ and $\hat{\rho}_{T+m}$ in order to bring the forecasts in line with the real world. In this case, the forecasts become:

$$_{T+m}\hat{Y}_{T+h} = (\hat{\rho}_{T+m})^{h-1} \left(Y_{T+m} - \sum f_i(\mathrm{X}_{i,T+m}) - \hat{\theta}_T\right) + \sum f_i(\mathrm{X}_{i,T+h}) + \hat{\theta}_{T+m}$$

And these forecasts can be used to generate the "audited" value of the firm. The approach adopts a *residual income* perspective (Fig. 11.3), where *value additivity* allows each component value flow to be defined independently, then *forecast, discounted* and *aggregated*. Discounting uses the own discount rate inherent in management's strategy and technology choice, neatly sidestepping the greatest problem in residual income approaches – assessing the opportunity cost of capital.

Figure 11.3 The Financial Dynamics approach to the valuation

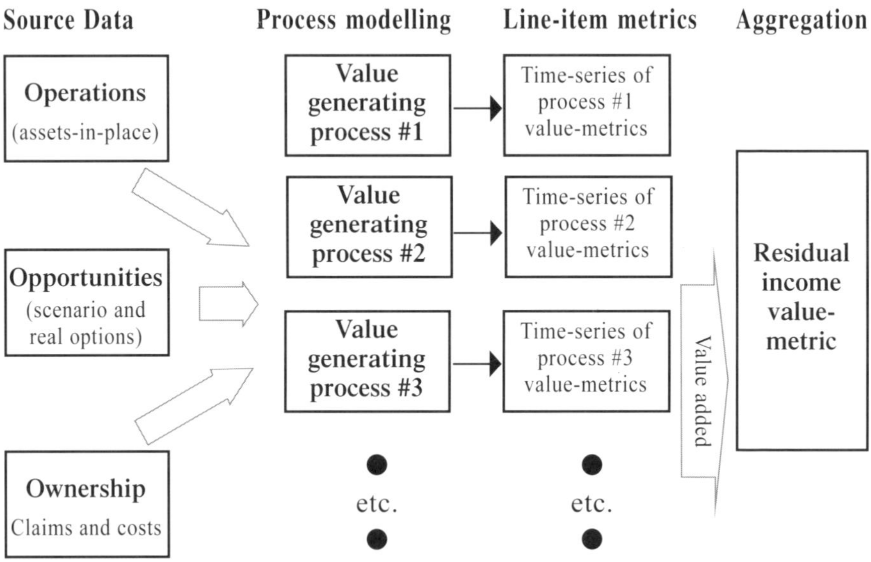

Transaction Reporting

The ten principles presented in *Valuing Technology* were motivated by the need for better reporting of firm operations. This section will show how Financial Dynamics methods incorporated into accounting systems can meet these needs. I will present the principles in the order in which they appeared in *Valuing Technology*, with a brief description of how a Financial Dynamics-based accounting system can satisfy or accommodate each principle.

- *Principle 1* (Sustainable value principle): "Real" and sustainable value is generated by managed processes.

- *Principle 2* ('Monte Carlo' principle): Speculative or holding gains from investments must be explicitly separated from value generated from operations.

Financial Dynamics graphs a firm's business processes in its value map, and defines these as the core activities[14] – the activities essential for value creation. Any other activities that may have in the past (or in the future) generated speculative or holding gains are segregated in the "environment" θ parameter of each value flow.

- *Principle 3* (Scalable matching principle): Processes that incur costs and generate revenues nonlinearly cannot be valued using a transaction approach; rather, these values must be generated with an integrated model of the particular set of operations which do not scale linearly.

The part of Pacioli's legacy that has caused the greatest confusion as business has become more complex involves an assumption of linear scaling of processes inherent in the *transaction approach*. Accounting's transaction approach assumes that most costs can be allocated over the units of output – simply divide the invoice cost by the number of widgets sold. This worked well for goods on consignment in Venetian ships, or for simple, cottage industry production. But where R&D knowledge, in-stream customization, network effects, branding, and other traits of a knowledge industry dominate product expense, simple allocations are inadequate.

Financial Dynamics addresses nonlinearities through the functions which map strategy drivers into value $\sum_i f_i(X_{i,T+1})$. The individual f_i may take any form that management believes correctly reflects the way value is accumulated – either linear or nonlinear. There is no demand on management to be 100% correct in the choice of $X_{i,T+1}$ or specification of f_i. Any error or misspecification simply shows up as variance in the estimator for the "environment" θ parameter.

- *Principle 4* (Technology acceleration principle): If technology acceleration and obsolescence is significantly large in a given class of costs, these costs should be classified as expenses and recognized in the current period.

"Cost" is a generic term that encompasses both expenses and asset investments. A cost represents a resource foregone (often a cash payment). If it generates value (that is, revenues) in the current period, it is an expense on the income statement; if value is generated only in the future, it is an asset on the balance sheet. Because all of our "dynamic" strategy drivers are *time periods*, and because the forecasts are over periods of time from T to $T+h$, the impact of dynamic scaling such as technology acceleration is seemlessly incorporated in Financial Dynamics.

- *Principle 5* (Value additivity principle): Processes contributing to firm value should be grouped based on the characteristic scaling of the process, valued according to the scalable matching principle, and added to compute a total valuation.

Financial Dynamics addresses nonlinearities through the functions f_i that management believes correctly reflects the way value is accumulated – either linear or nonlinear. The summation of individual components $\sum_i f_i(X_{i,T+1})$ provides computational flexibility and an intuitive perspective on the business models. Again, there is no demand on management to be 100% correct in the choice of $X_{i,T+1}$ or specification of f_i. Any error or misspecification simply shows up as variance in the estimator for the "environment" θ parameter.

- *Principle 6* (Core processes matching principle): Expenses generated by the core processes of a business should be matched (recognized in the same period) with the revenues generated by the products of those core processes; expenses generated by supporting processes should be expensed immediately as sunk costs.

The "core process matching principle" is met at the time that a *value metric* is selected for inclusion on the *value map*. Proper value metrics will ensure that this matching is accurate. Furthermore, the partitioning of particular value flows on to different parts of the value map ensures that there will be less confusion involved in properly matching. The clear statement of ownership and control in the value map also ensures that fiddles such as "dark swaps," common in the telecom bandwidth markets, are not allowed to distort account valuations.

- *Principle 7* (Error dispersion principle): Financial statements must report both account values and the estimated dispersion of error in the account value.

The "error dispersion principle" is perhaps the most important of these 10 principles for addressing potential misuse of statements. The knowledge economy has created enormous demand for risk management, as is evident in the dramatic growth of financial derivatives, and the difficulty of ascertaining risk. Much of the confusion arises because financial accounting reports account values, but nothing about the variance or error in those accounts (what the accounting profession loosely refers to as "materiality"). Fortunately, Financial Dynamics retains dispersion (that is, how variable transaction values are) from the very point of transaction reporting. Consider the basic estimation formula:

$$Y_t = \sum_i f_i(X_{i,t}) + \theta + u_t(\rho)$$

Then the u_t are assumed to follow a first-order autoregressive model $u_t = \rho u_{t-1} + \varepsilon_t$ where autocorrelation $|\rho| < 1$ and the ε_t's are independent with zero means and independent of the $X_{i,t}$'s. These ε_t's have variances (that is, dispersion statistics) which will be estimated from the regression (one each for value *P* and volume *Q*), and which provide a measure of "risk" in the "environment" parameter θ.

- *Principle 8* (Friedman principle): Accounting need only record those transactions that contribute information needed for the prediction of firm value.

Economist and Nobel laureate Milton Friedman[15] formally established the primary role of economic modeling for *prediction* in the early 1950s. Classical financial accounting, in contrast, defines a static model that accounts for historical contributions to wealth. This was fine for the Venetian merchant marine, English manorial accounts, and other simple trading businesses. It is completely inadequate for the knowledge economy. Financial Dynamics' major contribution is to bring accounting into line with contemporary perspectives on valuing firms. The central role of prediction of future value flows in Financial Dynamics ensures that the "Friedman principle" is satisfied.

- *Principle 9* ('Purchasing power' principle): Value is measured by "purchasing power" which is independent of nominal currency values.

Cash, because it is the *numéraire* commodity (i.e., the commodity used to compare the value of other commodities) fails to provide a stable measuring stick for value (neither do accounting measures, for that matter, though much of the complexity of accounting makes it more objective than pure cash measures). When most of the changes in the prices of the company's resources are in the same direction, the purchasing power of money can change dramatically. Conventional accounting statements are stated in nominal currency units (dollars, francs, lire, etc.), not in units of constant purchasing power. By splitting transaction flows into value and volume components, and by implicitly deflating the value component (or otherwise controlling for unstable monetary units) the resulting forecasts are agnostic with respect to currency fluctuations, and can be considered to accurately reflect underlying value.

- *Principle 10* ('Statistical' principle): Accounting transactions should be summarized and reported using statistical estimators with desirable properties.

Accounting statistics were state-of-the-art in 14th century Europe. But the dramatic evolution of statistics in the 20th century has provided many new tools for the informative reporting of business activity. The problem is profound in complex organizations, because the summarization of transactions into very broadly defined accounts tends to make their distributions highly kurtotic and thus difficult to use in forecasting or other analytical uses. Because a statistical model is central to Financial Dynamics, and the forecast estimators are unbiased, efficient, consistent, and sufficient for the data and drivers, this principle is met in Financial Dynamics.

THE END OF CERTAINTY

You may object that the calculations in Financial Dynamics are not precise, and, whereas this might work in econometrics, it is not precise enough for accounting, or the courtroom in which accountants increasingly find themselves. But accountants aren't precise; they just refuse to state the degree of their imprecision. In the accountant's vernacular, the accounts' imprecision is called "*materiality*." In court cases where audit materiality has been forced to be revealed, it has varied anywhere from a few percent of the account balance to 20–30% of the account balance. Of course, without explicit standards, the user of such statements hasn't the foggiest notion what the actual error in accounts might be.

Worse, it is likely that the upper limit of 30% materiality seen in court cases is less a limit, than a point at which litigants think less about suing for negligence, than for fraud. For example, in 2002, WorldCom admitted that they had overstated the prior 15 months of earnings by a total of US$7.1 billion, mainly through a system of so-called dark swaps – round-trip commodity trading of unused broadband. For a company that reported US$1.4 billion net income in 2001, it would be difficult to dismiss an error of several hundred percent as "*immaterial*." Of course, the users of WorldCom's financials had no idea of the magnitude of error, and as a result, WorldCom was able to raise US$12 billion in capital in 2001 from the sale of debt.[19]

The charade of precision confuses accountants in their conduct of audits. Inordinate amounts of effort and systems costs may be allocated toward ensuring that a particular account is tied out to invoices – say, for example, the property, plant, and equipment account – which will be completely washed out in errors of estimating depreciable life, or depreciation rate. We have outgrown a world of certainty in finance and accounting. This should not be surprising, as most of the sciences threw off the myth of a clockwork universe – be it in mathematics, physics, or biology – at some point in the 20th century. The dramatic growth of derivative financial markets and other means of managing risk are certain signs that accounting and finance are governed by the same laws as the rest of the universe.

The confusion of appropriate accounting measures with valuation goals – which vary greatly from situation to situation and constituency to constituency – and the resulting pressure to believe in a universal and unambiguous monetary measure for everything, tangible or intangible, tradable or not, have rippled perniciously through the profession and its academic study.

Many of us were puzzled by the takeover nearly 30 years ago in universities of academic study of accounting by a narrow orthodoxy enforced by the small group of academic journals that are available to report any influential academic accounting research. This orthodoxy has stifled the pursuit of meaningful research in accounting precisely at the time when the environment of the profession changed most profoundly, from a "gentlemen's club" that applied age-old traditions to audits, to an open market free-for-all that emphasized consulting;[20] from an axiomatic system that counted things, to a think tank that was required to measure and comment on all sorts of corporate activities. The revolution in accounting automation that started with the success of IBM's 360 line of computers in the 1960s transformed paper from an archival medium that provided permanent records for accounting transactions, to a display medium. In paper's place was substituted intangible 0s and 1s on computer media. Auditing's response was to develop software and methods that would extract these 0s and 1s to paper in order to continue their traditional ways. Professional auditing's attempts to come to terms with automation have led to some very odd and expensive hybrids.[21]

One prominent accounting academic described this narrow orthodoxy in stating that there are "only three models used in the study of accounting – the Brunswick Lens model, the Capital Asset Pricing Model (CAPM), and the Agency model."[22] The Brunswick Lens model studies the psychology of how accounting numbers are interpreted (often by bank loan officers and college sophomores, since these have traditionally been convenient sources of experimental data). The CAPM has been largely discredited as a description of stock market behavior, but accounting academics continue to use it as a way to glean whether accounting changes are noticed at all in the marketplace through what is known as an event study. Generally, the impact of accounting "events" has been small, and often precedes the actual change, indicating some exchange of inside information, casting further suspicion over the usefulness of accounting reports. The Agency model is perhaps the *most* puzzling of the standard academic accounting models, because it doesn't address accounting measurement at all, but rather the choice of "optimal" employment or insurance contracts in a firm. Usage of "optimal" in agency models doesn't necessarily relate directly to any concept used in accounting or audit practice.

The accounting profession has been poorly served and perhaps even misled by academic researchers who should have been the toolsmiths for the profession. While accounting academics have pursued research of questionable relevance to the profession, certain critical areas in auditing, computers, and the valuation of intangibles have been neglected. More fundamentally, the audit profession has failed to automate its activities along with the businesses they audit. Audits use many of the same techniques adopted in the 1930s, making them labor-intensive and risky. Some of these omissions directly led to the spectacular audit failures at Enron and WorldCom.

Contemporary accounting rests on archaic conceptions of economic activity and goals. Everything has a market (and market value); everything is infinitely divisible; risk is irrelevant. They are relics of the 14th century commodities trading model of Pacioli. Today, they provide a befuddled accounting profession with the security of a set of "natural laws" that can be called upon with impunity, and protects its practitioners from questioning lawyers and stockholders (although in today's environment even that is not guaranteed).

The knife that separates causal explanations of value from moral judgments cuts both ways – and accountants are in the middle of the debate. They can't help it. When the *Securities and Exchanges Act* of 1934 mandated audits for exchange-listed companies, it forced auditors to make a moral judgment that starts with the following statement:

> *"We believe these accounting statements to be fairly stated, in accordance with generally accepted accounting principles (GAAP) consistently applied..."*

This is the unqualified audit opinion that listed companies seek when they engage an auditor. In their perverse morality, accountants argue that they are merely technicians applying the age-old axioms of accounting. But we must see at least some intellectual laziness and moral cowardice in their failure to define "fair" (or, conversely, the "material error" that would make these statements "unfair"). The cloistering of accounting practice and moral reasoning in separate arenas lies behind the legal defense that management concealed their activities from accountants. Accountants have privileged access to corporate information; if they can't make a judgment, or if they can and it contains substantial risk, then this information needs to be available in the financial statements. Accountants cannot make both a moral declaration that the "statements are fairly presented" and at the same time hide behind the excuse of applying technical incantations to generate this moral judgment. This is management by superstition on the auditors' part; but audit incantations have grown suspect, as they have increasingly failed to deliver on their promises.

A mechanistic stance is important in assuring consistency, completeness, and other aspects important in an audit. "Fairness" is an idealization of firm activities that makes the morality game playable. Euclidean geometry requires idealizations such as infinite straight lines and perfect circles, and its deductions are sound and useful even though the world does not really have infinite straight lines of perfect circles. Similarly, moral judgments require idealizations such as "fairness" and "materiality" and "risk," and its conclusions can be sound and useful even though the real world does not really possess these characteristics innately. As long as there is no outright coercion or gross malfunction of reasoning, the approximation will be close enough to generate useful and defensible results.

The alternative is the muddle we witnessed after the collapse of the dot-coms: the savings of whole segments of society squandered on faulty deals and business models; undermining of confidence in capital markets and executives; and appropriation of the assets of the lower and middle classes by the most ruthless and derelict of the business elite. The world is right to heap moral opprobrium on accountants, though often the reasons may be wrong. Many of the accountants charged may have been good people trapped by uncompromising circumstances. Nonetheless, they got there by buying into a system of auditing and accounting that is antiquated, irrelevant, and misleading, but represented as up to the job it was applied to. This moral game continues today, and no band-aid will fix it. The old system needs discarding, with a new system to take its place – one which recognizes the moral imperative, risk, and consequences of auditing and accounting. One which is based on science, instead of tradition. One that helps its users make ethical as well as technical decisions.

BEYOND VALUATION

I hope the prior discussion has convinced you that a Financial Dynamics-based approach to firm accounting is feasible, and in fact desirable. Still, with the accounting profession being conservative, and with accounting systems being complex and mission critical, we are not likely to see a wholesale transformation of accounting in the near future. In fact, I would like for Financial Dynamics to insinuate itself into the accounting process as, perhaps, cost accounting did at the beginning of the 20th century; or as the managerial accounting and residual income approaches did in certain firms such as General Electric in the 1950s. But to justify the added data gathering and complexity of Financial Dynamics, firms would have to call upon it to solve some set of problems they face. The problem that is of primary importance on most corporate agendas is the increasing rapidity with which firms need to respond to the demands of external markets – both consumer and supplier.

Continuous-time accounting has been recommended for nearly 30 years now, but we seem little closer to agreeing on exactly how this would look than we were at the time Merton suggested this earlier. Financial Dynamics could provide the logical next step in the evolution of accounting, away from the current financial stewardship perspective and towards an accelerated real-time delivery of useful information. My final chapter discusses the way in which Financial Dynamics can be applied to the real-time decisions that confront managers.

---

[1] Quoted in Rupert Cornwell, "Shocked and Angry: The Prophet Whose Warnings over Wall Street Were Ignored," *The Independent*, July 1, 2002.

[2] Nanette Byrnes and David Henry, "Confused about Earnings?," *Business Week*, November 26, 2001, pp. 50–54. (Paul Volker is the former chairman of the Federal Reserve and member of the International Accounting Standards Board.)

[3] John Maynard Keynes *The General Theory of Employment, Interest and Money,* New York: Prometheus Books (Reprint edition, May 1997), Preface.

[4] George Lucas interview (with Bill Moyer), *Time* magazine, April 26, 1999.

[5] Quoted in "Badly in Need of Repair," *The Economist*, May 2, 2002, p. 67.

[6] Byrnes and Henry, op. cit.

[7] "Earnings: A Cleaner Look," *Business Week*, May 27, 2002. Nanette Byrnes and Mara Derhovanesian, "Standard & Poor's Gets Back to Basics With a New Measure of Profits," *Business Week*, May 27, 2002.

[8] Procter & Gamble was the most admired corporation in the 1950s, partly for the sophistication and innovativeness of its managerial accounting systems, (Indeed, its corporate structure was the model for departments at Harvard Business School, and subsequently most other business schools in the U.S.)

[9] The roots of this influence had much to do with the relative strength of the French economy, which was twice as large as the English, and the high regard in which French ideas were held after Louis XIV's generous funding of academic endeavors.

[10] Financial Accounting Standards Board, *Statement of Financial Accounting Concept No. 1: Objectives of Financial Reporting by Business Enterprises* (Stanford: November 1978).

[11] It may be more reasonable to use a Student *t*-distribution where sample sizes are small, or where the auditor believes transactions have a fat-tailed distribution, In this case, the problem would become $P(|\hat{\theta} - \theta| > t_{\alpha/2}\sigma) = \alpha$. The implications are the same as with the assumption of the normal distribution.

[12] Not only does the audit community adopt non-standard and confusing terminology, but the tables provided for calculations are presented without explanations of the formulas that have been used to generate them. This seems an egregious oversight when auditors regularly carry laptops loaded with spreadsheet processing software into the field, and could more easily simply run a macro to compute the table value. Such calculations are transparent, because the assumptions underlying them are explicit in the formula used in calculations. Two specific examples appear in the book *Audit Sampling (*AICPA, 2001). From Table 3.5 on p. 24, the table appears to compute a sample size as a proportion of (population size)$^{-0.75}$ though it is impossible from the accompanying narrative to figure out how they obtained this relationship; and Table 3.4 on p. 24 computes a sample size which is proportional to (expected population deviation rate)^2 – again without explanation. It is possible to find scenarios in which the two calculations would contradict each other. Without further specification from the auditors, though, it is difficult to ascertain what is being recommended.

[13] Calculated using Mathematica; the model assumes the annual account balance (that is, the value flow for the sales revenue account) equals the random variable $\tilde{u} = \tilde{v}_A \times \tilde{q}_A + \tilde{v}_B \times \tilde{q}_B$ where $\tilde{v}_\bullet$ is the value random variable for an individual transaction generated by the process and $\tilde{q}_\bullet$ is the number of transactions generated per year. Assuming that the random variables are stochastically independent (that is, that we are operating on a flat part of the demand curve) then the probability density function of $\tilde{u}$ the account random variable is computed in three steps:

1. $f_A(z) = \int \frac{1}{|x|} f_{V_A}(x)\, f_{Q_A}\left(z/x\right) dx$

2. $f_B(z) = \int \frac{1}{|x|} f_{V_B}(x)\, f_{Q_B}\left(z/x\right) dx$

3. $f_u(z) = \int f_A(x)\, f_B\left(z - x\right) dx$

where $f_{V_A}(x)$, f_{QA}, f_{VB}, and f_{QB} are all normal probability density functions with the parameters from the table.

[14] Gary Hamel and C. K. Prahalad have emphasized the concept of "core competencies" – things that must be done right for the success of the firm. The parallel concept of "core activities" used in this book includes all those critical activities that generate value while consuming scarce resources.

[15] M. Friedman, *Essays in Positive Economics* (Chicago: University of Chicago Press, 1953).

[16] Quoted in B. Nussbaum, "Can Trust Be Rebuilt?," *Business Week*, July 8, 2002, p. 36. Bush made this statement at an international meeting in Canada.

[17] Reported in The Economist, November 2, 2002, The Pitt and Webster Show, p. 71-2; further views can be found on the WWW, presented in Langevoort's briefs to the SEC, where he has served on a number of advisory committees.

[18] The stewardship function of accounting dominates accounting objectives as promulgated by the *Financial Accounting Standards Board (Financial Accounting Standards Board, Statement of Financial Accounting Concept No. 1: Objectives of Financial Reporting by Business Enterprises* (Stanford: November 1978)).

[19] "Curiouser and Curiouser," *South China Morning Post*, June 28, 2002, Business 2, p. 1.

[20] The elimination of various practices covered under the AICPA's Ethics Rules, section 501, in the 1980s allowed advertising, talent raids, and various other previously "unethical" practices in the interest of competition. Accounting firms suffered significant changes in thrust of the business, and in priorities of the partners which to some extent contributed to the excesses of the 1990s.

[21] For the interested, one high-profile example involving KPMG Peat Marwick is documented in Harvard Business School's Case Study "The Shadow Partner." KPMG's project married Macintosh home computers with massively parallel Connection Machines (primarily of use in weather forecasting) while attempting to spawn a data server (HBS Case Study 9-492-002, July 15, 1991, rev. October 5, 1995). Not surprisingly, the firm discovered that comparable performance could be obtained from Sun workstations at one-hundredth the cost, and there was even packaged software available. HBS Case Study 9-397-108 (July 11, 1997) documents progress a decade on.

[22] From a presentation by Cornell Professor Robert Libby at the University of Michigan School of Business, May 10, 1983.

CHAPTER 12

Managing the Real-Time Company

Accounting is being asked to do something it was never meant to do.
It is like a road system built for when cars did 30mph
being used by cars doing 60 –
and then we wonder why there are accidents.

– Robert Merton, professor,
Harvard Business School[1]

The "digital dashboard" metaphor has been used in the business press to describe an interactive cockpit display, somewhat like an airplane or automobile, which provides real-time feedback on the firm, its operations, and its competitive environment. Some software developers consciously employ this metaphor for software that integrates rapidly changing information from multiple sources into a unified display. Hewlett-Packard developed a tool for customizing Windows desktops called *Dashboard.* Microsoft, taking the cue, developed their *Digital Dashboard* tool to integrate Web-based data (for example, news, stock quotes, and so on) and internal data (for example, email, databases, and so on) into Office software, which offers customizable SQL Server database support. Recent press has been given to the idea of a general corporate or enterprise dashboard, or more specifically, a CIO or CEO dashboard as a vehicle for interactive "knowledge management."

The idea has been kicking around for decades. IBM's DL/1 (acquired from North American Rockwell) and CICS (acquired from Union Carbide) in the 1970s were designed to obtain, integrate, and disseminate information from a diverse set of corporate sources, controlling the "islands of data" mentality that fragmented files across many functional stovepipes. In the 1980s, at the pinnacle of IBM's power, they tried (and failed) to promote the model of unified corporate database, with real-time data access through IMS (the successor to DL/1, with integrated terminal access). In the 1990s, companies such as Lighthouse Computer Systems sought to promote Executive Information Systems, ostensibly to meet these same goals, but with a diversity of vendors and databases.

Unfortunately, the dashboard's "design by metaphor" provides scant actual design guidance, and could be accused of being gratuitous if not intellectually lazy. Pilots and drivers in a cockpit need to respond quickly to new information in a fast and constantly changing environment; real-time takes on a vivid significance when the outcome of poor decision making is one's own demise. Cockpit display in cars, airplanes, and other vehicles has been a point of much research and contention over the 20th century. And it is easy to see why: split seconds can make the difference between life and death. Cockpit feedback does not end with dials and displays, but is an essential part of controls such as the stick and steering wheel. Speed of response is gained by not overloading one system; for example, if tactile feedback can provide information about engine speed, turbulence, and so forth, then the visual system, which is already busy with a complex airplane display, should not be required also to process this information.

Businessmen have more time to make decisions than pilots; they also have considerably more information to process. The enduring appeal of real-time financial dashboard systems suggests that the demand for them is high, even if prior implementation has been disappointing. Still, there is a growing awareness by management of the need for centralized "knowledge management" though the meaning of that phrase is nebulous and mutable. Ideally, a dashboard should present real-time measurements of the environment so that correct decisions can be made in the shortest amount of time.

In general, the trend is toward cockpit displays that are designed to represent conceptual meaning in the perceptual properties of the display, along the lines of the "desktop" metaphor on Macintosh and Windows computers. Correct response to cockpit display feedback depends more on the interactions between pilots and cockpit displays than on the actions or decisions of an individual pilot working alone. A response involves the coordination of cockpit avionics, pilot actions and communications, and institutional procedures. The best-designed dashboard displays make appropriate information extractable with little cognitive effort.[1]

CAN YOU EVER HAVE TOO MUCH INFORMATION?

Better accounting does not necessarily mean more numbers – the interpretation that corporations and accountants have too often applied. *Quality* of information counts, not quantity. And quality implies that the appropriate measurements for the appropriate decisions are given to the appropriate users.

Gibboney Huske, an analyst at Credit Suisse First Boston covering Xerox, noted that the 1,000-page size of Xerox's 2001 annual report took her staff nearly a month to analyze. When a professional stock analyst has that much trouble understanding a company's financials, what hope is there for the average investor? In the wake of the investor cynicism of the post-Enron, post-WorldCom environment, almost every company is putting out longer, more complicated reports. Even those with far less complex finances are weighing in with 500-page statements, under the supposition that, when under assault, "more information is safer." The approach is reminiscent of a molested octopus – throw ink at the intruder, then change tack.

Can you ever have too much information? The short answer is "yes." The proper approach is not just more information, but more information presented in a way that can be easily adapted to the user's context. If the user is an investor, one that helps describe how well an investment in the company can be expected to perform; if a creditor, whether there will be cash flow problems; if a vendor, what the future market prospects are. Providing accountability is what accounting should be about.

The people who know about such things have recently been getting a bit more respect. Information retrieval is the science dedicated to finding the answers to topic-specific questions. It has long been associated with librarians, which has earned it a frumpy, dowdy countenance. In contrast, the traditional experts in business information search and retrieval have come from computer science, which dedicates whole journals to the organization of databases.

Two problems have plagued the database "supply-siders" since the days of DL/1: first has been predicting what users of a database ultimately want in advance of their questions; and second is wresting that data from the control of possessive departmental employees once the database has been designed. The latter effort, the one that might level out the "islands of data" on the corporate accounting landscape, has clashed with organizational politics that sabotaged many systems. By some accounts, over 90% of new systems projects fail because of political possessiveness over data.

We have very few examples of firms that have successfully installed company-wide access to data. Successes usually occur where an autocrat can order the organization of the firm around the data (for example, Dell Computer) or where the company systems are integrated with databases from the initial days of the corporation (for example, Cisco).

In contrast, since 1995 we have seen a number of successes in retrieving information from the most disaggregated set of databases in history – the World Wide Web. These solutions promote themselves under the rubric of "Web portals" with names like Yahoo, MSN, and Lycos. The most successful of these to date has been Google. Google uses the link structure of the Web as an indicator of an individual page's "*value*"; pages that are themselves "important" weigh more heavily and help to make other pages "important." This is an informal, automated way of valuing relevance and quality of information. In contrast, other portals such as Lycos use simple Boolean matching of search terms; or as Yahoo does, predict and index with manual labor, similar to a traditional DL/1 type of database. Database indexes can take up to 10–20% of the storage space in a corporate database, and are responsible for the significant overhead incurred in database management systems; so the choice is an important one. Promoters of the "digital dashboard" or its predecessors could learn something from Google's success in providing relevant responses to searches for Internet information.

WHAT IS REAL-TIME?

"Real –time" has been perhaps the most used and abused term in the business vernacular. The misuse of the 'digital dashboard' metaphor is just one example of the differing perceptions of 'real time' in business and other professions. Perhaps we can better define 'real-time' by reference to Webster's dictionary:

> *real-time* (adjective): Of or relating to a time frame imposed by external constraints. Real-time operations are those in which the machine's activities match the human perception of time or those in which computer operations proceed at the same rate as a physical or external process. Real-time operations are characteristic of transaction-processing systems, aircraft guidance systems, scientific applications, and other areas in which a computer must respond to situations as they occur (for example, animating a graphic in a flight simulator or making corrections based on measurements).

Usage of the term 'real-time,' when applied to information technology, has varied over the years. In the 1970s, "real-time online access" described the operation of the "dumb" terminals such as the mainstay DEC VT100. Or for example, the description of real-time (only if you have a lot of time) access via wireless 3G telephones. But the great revolutions in information technology have always taken place in the back office, too often unnoticed by chroniclers, but of importance to businessmen.

"Real time" describes the full potential of what information technologies could ultimately do, according to Andrew Odlyzko, director of the Digital Technology Center at the University of Minnesota: drive the lags and latencies out of the economy and make it much more efficient. Odlyzko predicts that real-time technology will prove as important for speeding up the information flow in business as was the telegraph, invented in 1837.

Real-time companies have become the new darling of venture capitalists. Whether or not this will prove a passing fad, or an enduring transformation, it is important to investigate whether their claims that highly complex enterprise software, based on advanced mathematics, can yield real savings if deployed and used correctly. Real-time firms are seen as the next wave of successful companies. What is more, the current recession will probably speed up the adoption of new real-time products and services. Many of these do not require big investments. Instead, they make existing computer systems more effective, promising a quick return. General Electric estimates that its digitization efforts saved it US$1.6 billion in 2001. Firms that have spent billions on IT without any guarantee that their bottom line would benefit are desperate for such savings. Most important, perhaps, the concept of real time is more useful than the somewhat artificial distinction between the "old" and the "new" economy.

When engineers designed NASA's space shuttle – ostensibly an airplane that would be able to maneuver in outer space – they presumed that pilots would be in control the entire flight (just like in *Star Wars*). But, because the shuttle re-enters the atmosphere at about 17,000 mph (28,000 kph), at a 40-degree attitude (a normal commercial airliner has about a 3-degree attitude), human control simply would not be fast enough to execute the intricate series of S-shaped, banking turns needed to slow its descent speed for the final approach to the runway. Only at 25 miles (40 kilometers) out do the shuttle's landing computers give up control to the commander. Without automated control, landing the shuttle would be suicide. Even commercial pilots rely on automated pilots to relieve the tedium of long flights, saving their attention and expertise for takeoffs, landings, and emergencies.

As firms grow more complex, and real-time monitoring more common, we can expect the traditional concepts of "command-and-control" and much-vaunted "leadership," to similarly give way to automated systems – if only for survival in the face of competition. Strategy today relies on a rapid response to evolving environments. Even where managers can exert a "hands-on" leadership, it may not be optimal to make all decisions at human speed. Indeed, on manufacturing's supply side, electronic data interchange and enterprise resource planning have taken people out of the loop, making inventory control fast, cheap, and hyper-efficient in a way that not even the greatest leaders could.

An introspective look at the character of leadership would reveal that it often is a call to the personal capabilities of a particularly charismatic, smart, or prescient individual to fill the gaps left by incomplete, flawed, or obsolete corporate policy and strategy, to bridge the gap between formal and informal systems, and to provide a unified "vision" for the activities of the firm. These are often needed; the question I raise is whether – as business speeds up – we can find individuals who can meet this challenge in a faster, more complex business world. My guess would be that individuals who can manage ultra-fast, ultra-complex 21st-century knowledge firms will eventually become as rare as those who can manually fly the space shuttle.

Asian managers have grown fond of citing the virtues of "*guanxi*" and "crony capitalism" – the presumed economic guidance of developing economies by an educated and inbred coterie of businessmen-politicians. In contrast, the U.S. and Europe developed over the past few centuries under political systems that held theory, principles, and abstract ideas above the influence of individual men. Not so in Asia, where nearly every growth economy has only within a generation cast off colonial rule, dictatorship, or absolute monarchy. In such societies, *guanxi* is indispensable. It is also increasingly inefficient, as a fixed group of associates over a long time span are more likely to be gatekeepers and tax farmers than contributors to productivity.

Leadership is personal; it is human. Efficient decision-making should not be bogged down by the chain of command – it needs individual decision-making at its source, based on guidelines, policy, standards, principles, and so forth. The balance of power between man and machine has been changing for two centuries, ever since the Industrial Revolution. We are just beginning to witness a parallel change in knowledge-intensive industries; a change which over the next century is likely to be even more profound as our world evolves to one dominated by leisure and creativity, rather than tedious and structured work (at which machines are exceptionally capable).

One of the great debates of the late 20th century was whether or not humans were actually needed for space exploration. The dénouement of the debate was that, for the most part, the extreme and hazardous conditions of space are best suited for purpose-built machines. Humans should stay on the ground and concentrate on strategy, planning, and interpretation of data. (A parallel debate is raging in the equally extreme and hazardous realm of warfare, where humans still, decidedly, have a role to play.) In business, which is less physically hazardous, the machine will win, but expect the debate to carry on much longer.

FINANCIAL DYNAMICS AND CONTINUOUS-TIME ACCOUNTING

Up to this point, we have assumed a fundamental time unit of one solar year both for traditional accounting, as well as the Financial Dynamics fiscal archeology based on numbers provided by traditional accounting systems. But this is arbitrary, and we could theoretically exchange it for a week, a day, or an hour. Such a change can be easily incorporated into Financial Dynamics through the 'auditing' formula presented in Chapter 10. Management, first, determines the minimum time that interval constitutes 'real-time.' This minimum interval is the 'granularity' of a Financial Dynamics based continuous-time accounting system. Financial Dynamics 'audits' the financial statements by updating estimates of environment and strategy lag from the updated database after each 'granularity' interval in order to bring the forecasts in line with the real world

We would presume that control and error correction occurs at the point of transaction entry, as has been the case since computers came to dominate accounting three decades ago. For efficiency reasons, it may still be prudent to batch transactions, if only because errors will require some human intervention. Assume that transactions are batched daily, and that "continuous-time accounting" is accounting that is updated daily. Certainly, there is demand for such accounting; for example, retailers have since the early 20th century produced daily "flash sales" detailing revenue by product line. "Flash sales" reports allowed retail buyers and shop clerks to modify purchase orders or displays based on customer trends. Wal-Mart has come to control 60% of U.S. retailing through extensive use of their own computer-intensive sorts of "flash sales" reports.

What is different today? First, computers give us the capability of creating reports on-the-fly, and tailored to the specific needs of a decision-maker in a specific context. Or we can have computer algorithms – intelligent agents, so to speak – that can alert managers to information when it becomes available. If you believe Alan Turing, any procedure reducible to a finite number of explicit steps can be automated – basically, if a computer could do all mathematical operations, it could also do anything a person can do.

Second, computers and databases allow us to generate much more sophisticated statistics describing transactions than was possible with spreadsheet and ledger. For example, rather than the dollar amount of accounts receivable on a balance sheet, we might like the mean, standard deviation, and count of individual accounts receivable transactions (from which we could recompute the amount if desired). Proper statistics, rather than mere account balances, provide analysts and investors with considerably more flexibility in generating analyses.

Finally, with the networked computer, investors and managers are no longer constrained to preprinted paper reporting. These new principles provide a basis for *dynamically* preparing *ad hoc* reports on-the-fly, and for circulating new information based on personal profiles and the assessment of computer algorithms. They provide an evolutionary path from the static accounting reports of the English manorial system, toward a new *science of accounting*.

EVOLUTION

Accounting has come a long way since the days of King Croesus. Minting of coins in ancient Lydia had the unanticipated side-effect of popularizing many previously exclusive realms of knowledge. An entrenched priesthood dictating the allocation of wealth through their mathematical magic gave way to commercial democracy of merchants who could count, add and were literate in previously exclusive sciences. History illustrates the varied set of disciplines that over time owed their evolution to accounting – mathematics, language, democracy and economics. For all its past contributions, though, modern accounting has become mired in a conservatism that threatens to make it irrelevant to a world that is evolving around it. The most important achievement of this book will be, I hope, to help us escape from the present medieval perspective on wealth and value formation. The components I have used in constructing Financial Dynamics are the simplest set which can satisfy three requirements: (1) forecast future value based on known history; (2) explicitly incorporate managerial strategy into models of firm value creation; and (3) incorporate global technology developments into the dynamics of firm value creation. I've worked hard to keep it simple.

John Gall[1] observed that "A complex system that works is invariably found to have evolved from a simple system that worked." Successfully building the simplest model that does the job has a long tradition in science. Claudius Ptolemy's *Almagest*, the classic compendium of science in the ancient and medieval world, recommended that "the scientist choose the simplest model consistent with the facts," an admonition reiterated as Occam's Razor in medieval science. There is ample room for refinement of the techniques presented in this book, and many tools exist in economic, statistical, and other academic literature that can extend the scope and utility of Financial Dynamics. Financial Dynamics is only a beginning; one that with luck will bootstrap accounting into a future where tangible things will take a backseat to the mind – the future knowledge economy in which a partnership of mind and machine defines all that is valuable in this world.

---

[1] John Gall, *Systemantics: The Underground Text of Systems Lore* (Walker, MN: General Systemantics Press, 1986).

[2] Distributed cognition theory (for example, see E. Hutchins, *Cognition in the Wild*.(Cambridge Mass: MIT Press, 1995) provides a framework for analyzing systems in terms of how representations are created, configured, transformed, and propagated through a system in the context of an activity. The inventory of representations in the system include speech and movements, cockpit (displays, switches, dials), checklists, and procedures. These are discussed more extensively in A. F. Stokes and C. D. Wickens, "Aviation Displays," in E. Wiener and D. Nagel (eds.), *Human Factors in Aviation* (San Diego, CA: Academic Press, 1988), p. 387–422; and J. Zhang, "Distributed Representation as a Principle for the Analysis of Cockpit Information Displays," *The International Journal of Aviation Psychology*, 7(2), 1997, p. 105–21.

Index

A

B

C

D

E

F

G

H

I

J

K

L

M

N

O

P

Q

R

S

T

U

V

W

X

Y